Air Valves:
Air-Release, Air/Vacuum and Combination

Second Edition

*The AWWA Standards Committee on Air Valves and the AWWA Standards Council
have reaffirmed this manual as current on November 21, 2025.*

American Water Works Association

Manual of Water Supply Practices—M51, Second Edition

Air Valves: Air-Release, Air/Vacuum and Combination

Disclaimer

If you find errors in this manual, please email books@awwa.org. Possible errata will be posted at www.awwa.org/resources-tools/resource.development.groups/manuals-program.aspx.

Managing Editor-Book Products: Melissa Valentine
Production: Janice Benight
Cover design: Melanie Yamamoto

Library of Congress Cataloging-in-Publication Data

Names: Ballun, John V., author. | American Water Works Association, issuing
   body.
Title: Air-release, air/vacuum, and combination air valves / by John V.
   Ballun.
Other titles: AWWA manual ; M51.
Description: Second edition. | Denver, CO : American Water Works Association,
   [2016] | Series: Manual of water supply practices ; M51 | Includes
   bibliographical references and index.
Identifiers: LCCN 2016019688 | ISBN 9781625761767
Subjects: LCSH: Water-pipes--Valves | Air valves.
Classification: LCC TD491 .B34 2016 | DDC 628.1/5--dc23
LC record available at https://lccn.loc.gov/2016019688

Printed in the United States of America

ISBN   978-1-62576-176-7   eISBN-13   978-1-61300-387-9

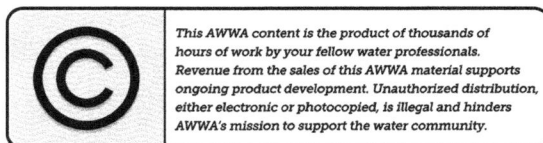

American Water Works Association

American Water Works Association
6666 West Quincy Avenue
Denver, CO 80235-3098
awwa.org

Contents

Figures

This page intentionally blank.

Tables

This page intentionally blank.

Preface

This manual is a guide for selecting, sizing, locating, and installing air valves in water and wastewater applications. Water includes raw water, potable water, and reclaimed wastewater that has been treated. Wastewater is a combination of used liquid and liquid-carried waste from residences, commercial buildings, industrial plants, and institutions, together with any groundwater infiltration, surface water, and stormwater that may enter the collection system.

The manual is a discussion of recommended practice, not an American Water Works Association (AWWA) standard. It provides guidance on generally available methods and capacity information. Questions about specific situations or applicability of specific valves should be directed to the manufacturer or supplier.

Information contained in this manual is useful to operators, technicians, and engineers for gaining a basic understanding of the use and application of air valves. There are many special liquid piping systems that are beyond the scope of the methodology given in this manual and may require special tools such as computer programs for analysis of hydraulic transients. The valve capacity information is generic information. Actual capacity charts of the intended manufacturer's valve should be consulted before making the final selection of valve size and options. The manual provides information only on the air valve types listed in ANSI/AWWA Standard C512, latest edition, including the following:

- Air-release valve
- Air/vacuum valve
- Combination air valve

Vacuum breakers, slow-closing devices, and throttling devices are also discussed in this manual. Other sources of information should be consulted for the use and application of these devices.

This second edition includes new or revised information pertaining to wastewater applications, penstocks, slow-closing devices, throttling devices, and vault products for freeze and flood protection. Also, new and alternate air valve sizing methodologies were added for partial rupture gravity flow and air-release valves.

Manufacturers graciously provided valve illustrations and other documentation. AWWA does not endorse any manufacturer's products, and the names of the manufacturers have been removed from the material provided.

This page intentionally blank.

Acknowledgments

The AWWA Standards Subcommittee on the Air Valve Manual M51, which developed this manual, had the following personnel at the time of approval.

John V. Ballun, *Chair*

D. *Alexander*, Cla-Val Company, Costa Mesa, Calif.	(AWWA)
J.V. *Ballun*, Val-Matic Valve & Manufacturing Corporation, Elmhurst, Ill.	(AWWA)
J.J. *Cusack Jr.*, Bryant Associates, Braintree, Mass.	(AWWA)
R. *DiLorenzo*, Mundelein, Ill.	(AWWA)
D.M. *Flancher*,* Standards Engineer Liaison, AWWA, Denver, Colo.	(AWWA)
F.H. *Hanson*, Albert A. Webb Associates, Riverside, Calif.	(AWWA)
R. *Kadava*, Black & Veatch, Kansas City, Mo.	(AWWA)
L. *Larson*, DeZURIK-APCO-Hilton Inc., Sartell, Minn.	(AWWA)
B.J. *Lewis*, Crispin Multiplex Manufacturing Company, Berwick, Pa.	(AWWA)
M. *MacConnell*, Metro Vancouver, Burnaby, B.C., Canada	(AWWA)
D.L. *McPherson*, HDR Engineering Inc., Charlotte, N.C.	(AWWA)
W.J. *Nicholl*, GA Industries, LLC, Cranberry Township, Pa.	(AWWA)
T. *O'Shea*, DeZURIK-APCO-Hilton Inc., Schaumberg, Ill.	(AWWA)
L.J. *Ruffin*, Ruffin Companies, Orlando, Fla.	(AWWA)
J.H. *Wilber*, American AVK, Littleton, Colo.	(AWWA)
N. *Zloczower*, A.R.I. Flow Control Accessories, Israel	(AWWA)

The AWWA Standards Committee on Air Valves, which developed and approved this manual, had the following personnel at the time of approval:

Miles E. Wollam, *Chair*

General Interest Members

A. *Ali*, ADA Consulting Ltd., Surrey, B.C., Canada	(AWWA)
J.H. *Bambei Jr.*, Denver Water Department, Denver, Colo.	(AWWA)
D.E. *Barr*, ms consultants inc., Columbus, Ohio	(AWWA)
J.J. *Cusack Jr.*, Bryant Associates, Braintree, Mass.	(AWWA)
R. *DiLorenzo*, Mundelein, Ill.	(AWWA)
D.M. *Flancher*,* Standards Engineer Liaison, AWWA, Denver, Colo.	(AWWA)
R.G. *Fuller*,† HDR Engineering Inc., Denver, Colo.	(AWWA)
F.H. *Hanson*, Albert A. Webb Associates, Riverside, Calif.	(AWWA)
D.L. *McPherson*, HDR Engineering Inc., Charlotte, N.C.	(AWWA)

---

* Liaison, non-voting
† Alternate

W.L. Meinholz, AB&H, A Donahue Group, Chicago, Ill.   (AWWA)

J.W. Snead II, JQ Infrastructure, Dallas, Texas   (AWWA)

T.J. Stolinski Jr., Black & Veatch Corporation, Kansas City, Mo.   (AWWA)

M. Stuhr,* Standards Council Liaison, City of Portland,
    Portland, Ore.   (AWWA)

R.J. Wahanik, Hystras, Wyommissing, Pa.   (AWWA)

R.A. Ward, Tighe & Bind, Westfield, Mass.   (AWWA)

M.E. Wollam, MWH Global, Pasadena, Calif.   (AWWA)

Producer Members

D. Alexander, Cla-Val Company, Costa Mesa, Calif.   (AWWA)

J.V. Ballun, Val-Matic Valve & Manufacturing Corporation,
    Elmhurst, Ill.   (AWWA)

L. Larson,† DeZURIK-APCO-Hilton Inc., Sartell, Minn.   (AWWA)

B.J. Lewis, Crispin Multiplex Manufacturing Company,
    Berwick, Pa.   (AWWA)

J.D. Milroy, Henry Pratt Company, Aurora, Ill.   (AWWA)

W.J. Nicholl, GA Industries, LLC, Cranberry Township, Pa.   (AWWA)

T. O'Shea, DeZURIK-APCO-Hilton Inc., Schaumberg, Ill.   (AWWA)

J.M. Radtke, Aqua-Dynamic Systems Inc., Wilkes-Barre, Pa.   (AWWA)

K. Sorenson,† A.R.I. Flow Control Accessories, South Jordan, Utah (AWWA)

J.H. Wilber, American AVK, Littleton, Colo.   (AWWA)

N. Zloczower, A.R.I. Flow Control Accessories, Israel   (AWWA)

User Members

L. Aguiar, Miami Dade Water and Sewer Department,
    Miami, Fla.   (AWWA)

R. Crum, City of Titusville, Titusville, Fla.   (AWWA)

N.E. Gronlund, East Bay Municipal Utility District,
    Oakland, Calif.   (AWWA)

M. MacConnell, Metro Vancouver, Burnaby, B.C., Canada   (AWWA)

P. Ries, Denver Water Department, Denver, Colo.   (AWWA)

B. Schade, WaterOne, Lenexa, Kan.   (AWWA)

M.I. Schwartz, Loudoun Water, Ashburn, Va.   (AWWA)

J.A. Wilke, Seattle Public Utilities, Seattle, Wash.   (AWWA)

---

† Alternate
* Liaison

Chapter **1**

Introduction

Air valves are hydromechanical devices designed to automatically release air and wastewater gases or admit air during the filling, draining, or operation of liquid piping systems for water and wastewater services. The safe and efficient operation of a liquid piping system is dependent on the continual removal of air and wastewater gases from the liquid piping system. This chapter includes an explanation of the effects of air and wastewater gases and their sources in liquid piping systems.

OCCURRENCE AND EFFECT OF AIR AND WASTEWATER GASES IN LIQUID PIPING SYSTEMS

Water contains approximately 2 percent dissolved air or gas by volume at standard conditions (14.7 psia [101 kPa absolute] and 60°F [16°C]) (Dean 1992) but can contain more, depending on the liquid pressure and temperature within the liquid piping system.

Wastewater systems can also contain more undissolved air and wastewater gases due to the decomposition of materials in the wastewater. Dissolved air and wastewater gases can come out of solution in pumps and in different locations along the liquid piping system where turbulence, hydraulic jumps, and other pressure variation phenomena occur. Once out of solution, air and wastewater gases will not readily dissolve and will collect in pockets at high points along the liquid piping system.

Air and wastewater gases come out of solution in a liquid piping system due to low-pressure zones created by partially open valves, cascading flow in a partially filled pipe, variations in flow velocity caused by changing pipe diameters or slopes, and changes in pipe elevation. Entrained air that reaches water service connections may be detrimental to the customer's water systems.

An air and wastewater gas pocket may reduce the flow of liquid in a liquid piping system by reducing the cross-sectional flow area of the pipe, and if the volume of the air and wastewater gas pocket is sufficient, complete binding of the liquid piping system is possible, stopping the flow of liquid (Karassik et al. 2007).

The velocity of the flow of liquid past an enlarging pocket of air and wastewater gases may only be sufficient to carry part of the pocket of air and wastewater gases downstream unless the liquid velocity is greater than the critical velocity for transporting air and wastewater gases in that particular pipe diameter (Escarameia et al. 2005). The velocity needed to scour a pocket of air and wastewater gases in larger piping systems (e.g., 24 in. [610 mm]) may be as high as 7.1 ft/sec [2.2 m/sec] at a 5 percent slope as shown in Table 1-1 (Jones et al. 2008). Although the flow velocity of the liquid may prevent the liquid piping system from complete air and wastewater gas binding, the pockets of air and wastewater gases will increase head loss in the liquid piping system (Edmunds 1979). As shown in Figure 1-1, a pocket of air and wastewater gas can reduce the flow in the pipe to d and create head loss equal to H_L due to the restricted cross section. Additional head loss in a liquid piping system decreases the flow of liquid and increases power consumption required to pump the liquid. Pockets of air and wastewater gases in a liquid piping system are difficult to detect and will reduce the liquid piping system's overall efficiency.

Pockets or air and wastewater gases may also contribute to water hammer problems, pipe breaks, system noise, and pipe corrosion—especially hydrogen sulfide corrosion—and can cause erratic operation of control valves, meters, and equipment. Studies have shown that small pockets of air and wastewater gases in certain locations along the system can cause transients and surge and/or intensify transients and surges, including downsurges (Pozos-Estrada 2007). However, temporary pockets of air and wastewater gases may be needed in special circumstances to prevent vacuum conditions in a liquid piping system after pump outages or line breaks. Vacuum conditions should be avoided as they may result in collapse and/or deformation of thin-walled pipe. Finally, on system applications, in locations where liquid column separations and returns may occur, a vacuum breaker with air-release valve or an air valve with restricted outflow (slow-closing device or throttling device) should be considered.

Table 1-1   Velocities required to scour air and wastewater gases from pipelines

Pipe Size	Scouring Velocities by Pipe Size and Slope				
	Velocity, ft/sec (*m/sec*)				
	Negative Slope				
in. *(mm)*	0° (0%)	2.9° (5%)	14° (25%)	45° (50%)	90° (vertical)
4 (100)	2.7 (0.8)	2.9 (0.9)	3.1 (0.9)	3.4 (1.0)	3.5 (1.1)
8 (200)	3.8 (1.2)	4.1 (1.2)	4.4 (1.3)	4.8 (1.5)	5.0 (1.5)
12 (300)	4.7 (1.4)	5.0 (1.5)	5.4 (1.6)	5.9 (1.8)	6.1 (1.9)
24 (600)	6.6 (2.0)	7.1 (2.2)	7.6 (2.3)	8.3 (2.5)	8.6 (2.6)
36 (900)	8.1 (2.5)	8.7 (2.7)	9.3 (2.8)	10.2 (3.1)	10.6 (3.2)

Source: Jones et al. 2008.

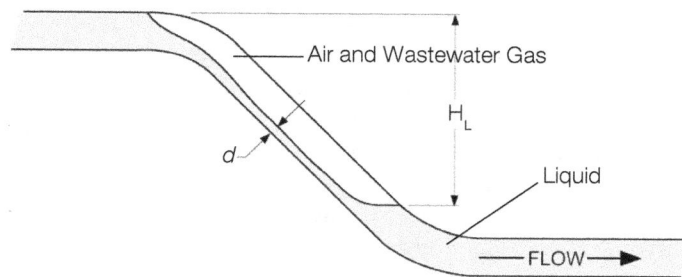

Figure 1-1   Air and wastewater gas pocket in a pipeline

SOURCES OF AIR AND WASTEWATER GASES IN LIQUID PIPING SYSTEMS

In addition to air and wastewater gases coming out of solution, air may enter liquid piping systems at leaky joints where the pressure within the liquid piping system falls below atmospheric pressure. These conditions exist in the vortex at the pump suction, at pump glands where negative pressure occurs, and at all locations where the pipe elevation is above the hydraulic grade line.

Air may enter liquid piping systems through air/vacuum and combination air valves following complete pump shutdown, through the orifices of air-release valves installed in locations where the pressure is less than atmospheric, and through pump suction pipes or inlet structures that are not properly designed to prevent vortexing. Finally, most vertical turbine and well pumps start with air and wastewater gases in the pump column as shown in Figure 1-2, which may pass by the check valve and flow into the liquid piping system with every pump start.

Air and wastewater gases entrainment is much greater in wastewater force main systems than in other pumped liquid transmission systems owing to their unique design and operational characteristics. Lift stations with wet wells or other sewage collection basins are a major source of entrained air and wastewater gases induced by plunging jets of sewage and by vortices of air and wastewater gases sucked into the pump. Because of the cyclic operation of force main systems, sections of the force mains empty out at the end of each pumping cycle, drawing air and wastewater gases into pipes. At the entrance to sewage lift stations, air and wastewater gases are entrained from plunging jets of sewage.

AIR AND WASTEWATER GAS POCKET BEHAVIOR IN PIPELINES

Four major factors influence entrained air and wastewater gas behavior in liquid piping systems: buoyancy, velocity, drag, and equilibrium in surface tension between the liquid, air and wastewater gases, and the pipe wall. These factors, together with air and wastewater gas pocket size and concentration, influence the tendency of bubbles to aggregate and increase in size and determine the direction of their movement either with or opposite to the direction of liquid flow. These factors also affect the entrained air and wastewater gases pockets' influence on liquid flow capacity, head loss, and energy consumption. In rising pipe sections and when there is no flow in the pipeline, buoyancy will force air and wastewater gas pockets of all sizes and shapes to travel to peaks or high points along the liquid pipeline. At downsloping and level pipe sections, when buoyancy exceeds drag, the pockets will travel upward in the opposite direction to the flow. When drag exceeds buoyancy, the pockets will travel in the direction of liquid flow. Large air and wastewater gas pockets traveling in opposite direction to the liquid flow often break up in flow due to buoyancy, resulting in smaller air and wastewater gas pockets, including bubbles, changing direction and being dragged in the direction of the liquid flow with the larger air and wastewater gas pockets or continuing to travel upstream against the flow. Pockets of air and wastewater gases traveling with the liquid stream also break up into smaller air and wastewater gas pockets and bubbles that disperse in the liquid stream, traveling in different velocities. In all these cases, the air and wastewater gas pocket's movement disturbs the flow where drag and turbulence increase head losses, resulting in decreased flow capacity and increased energy consumption (Lubbers and Clemens 2005).

Figure 1-2   An air valve installed on the outlet of a pump and upstream of the check valve

REFERENCES

Dean, J.A. 1992. *Lange's Handbook of Chemistry,* 14th ed. New York: McGraw-Hill.

Edmunds, R.C. 1979. Air Binding in Pipes. *Jour. AWWA,* 71(5): 272–277.

Escarameia, M., C. Dabrowski, C. Gahan, and C. Lauchlan. 2005. *Experimental and Numerical Studies on Movement of Air in Water Pipelines,* Report SR 661, Release 3.0. Wallingford, Oxfordshire, UK: HR Wallingford Ltd.

Jones, G.M., R.L. Sanks, G. Tchobanoglous, and B.E. Bosserman II, eds. 2008. *Pumping Station Design,* rev. 3rd ed. Boston: Butterworth-Heinemann.

Karassik, I.J., J.P. Messina, P. Cooper, and C.C. Heald. 2008. *Pump Handbook,* 4th ed. New York: McGraw-Hill.

Lubbers, C.L., and F.H.L.R. Clemens. 2005. Capacity Reduction Caused by Air Intake at Wastewater Pumping Stations. In *Proc. Water and Wastewater Pumping Stations Conference, April.* Cranfield, UK: University of Cranfield.

Pozos-Estrada, O. 2007. Investigation on the Effects of Entrained Air in Pipelines, Dissertation Volume 158. Stuttgart, Germany: Institute of Hydraulic Engineering Faculty of Civil and Environmental Engineering, Engineering and Allied Operations, Institut für Wasserbau der Universität Stuttgart.

This page intentionally blank.

Chapter **2**

Types of Air Valves

This chapter discusses the three basic types of air valves used in the water and wastewater industry that are included in ANSI/AWWA C512, latest edition, Air Valves: Air-Release, Air/Vacuum and Combination for Water and Wastewater Service. Water includes raw water, potable water, and reclaimed wastewater that has been treated. Wastewater is a combination of the used liquid and liquid-carried waste from residences, commercial buildings, industrial plants, and institutions, together with any groundwater infiltration, surface water, and stormwater that may enter the collection system. Air valves for wastewater service meet special design requirements, which may include enlarged inlet connections of at least 2 in. (50 mm), elongated bodies with sloped bottoms, bronze or stainless-steel trim, and interior coatings (ANSI/AWWA C512).

AIR-RELEASE VALVES

Air-release valves, also called *small-orifice valves,* are hydromechanical devices designed to automatically vent small pockets of accumulated air and wastewater gases as they accumulate within a full and pressurized liquid piping system. It is important to note that the term *air-release valve* is sometimes used as a generic term for *air valve* but in this document and in ANSI/AWWA C512, the term *air-release valve* relates to only the small-orifice–type valve. Examples of air-release valve mechanisms are shown in Figure 2-1. Air-release valves are characterized by outlet orifices that are much smaller than the inlet connection or connecting pipe size. Orifice sizes generally fall between an area equivalent to a circle with a diameter of 1/16 in. (1.6 mm) and 1 in. (25 mm), while the inlet connections can range from 1/2 in. (13 mm) to 6 in. (150 mm) in diameter. The wastewater version features an elongated body to prevent solids from collecting in the lever mechanism. Flush and drain ports are also provided to assist in cleaning or backwashing the valve.

When installed, the valve is normally open and will vent air and wastewater gases through the orifice. As liquid enters the valve, the float rises, closing the orifice. When air and wastewater gases, which have accumulated in the liquid piping system, enter the

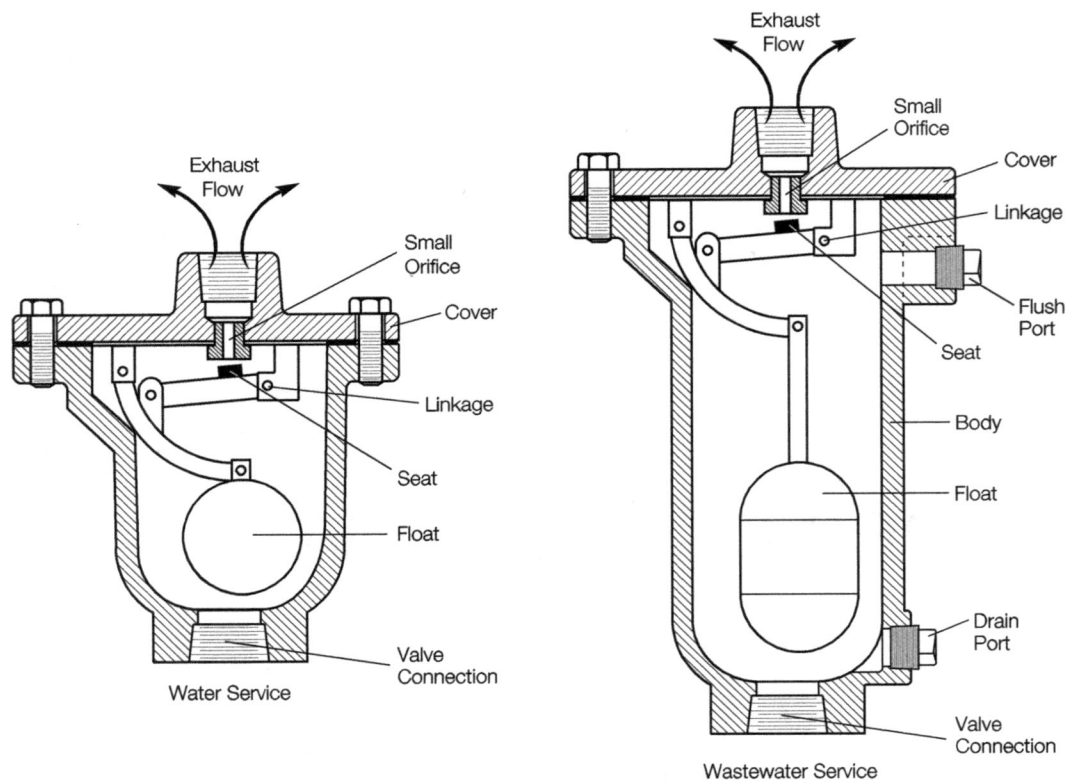

Figure 2-1   Air-release valves

valve, they displace the liquid, causing the float to drop and allowing the air and wastewater gases to vent through the orifice. An air-release valve designed with the proper float weight and proper sealing mechanism design, such as rolling seals or leverage mechanism, will allow the valve to open at any pressure up to the maximum operating pressure of the valve as designated by the manufacturer.

AIR/VACUUM VALVES

Air/vacuum valves, also called *large-orifice valves,* are direct-acting float-operated or diaphragm-operated hydromechanical devices designed to automatically vent or admit large volumes of air and wastewater gases during the filling or draining of a liquid piping system. This valve will open to relieve negative pressures and when closed will not reopen to vent air and wastewater gases when the piping system is full and operating at a pressure exceeding atmospheric pressure. The negative pressure may be due to column separation, system draining, controlled and uncontrolled pump shutdown, check valve or isolating valve closure, or a break in the liquid piping system and so on. Examples of air/vacuum valves are shown in Figure 2-2. The "inlet" of the valve is connected to the liquid piping system. The "outlet" may or may not be equipped with a hood or exhaust piping and is used for the inward flow of air and the outward flow of air and wastewater gases. The wastewater version features an elongated body to prevent solids from collecting in the large orifice with flush and drain ports to assist in cleaning or backwashing the valve.

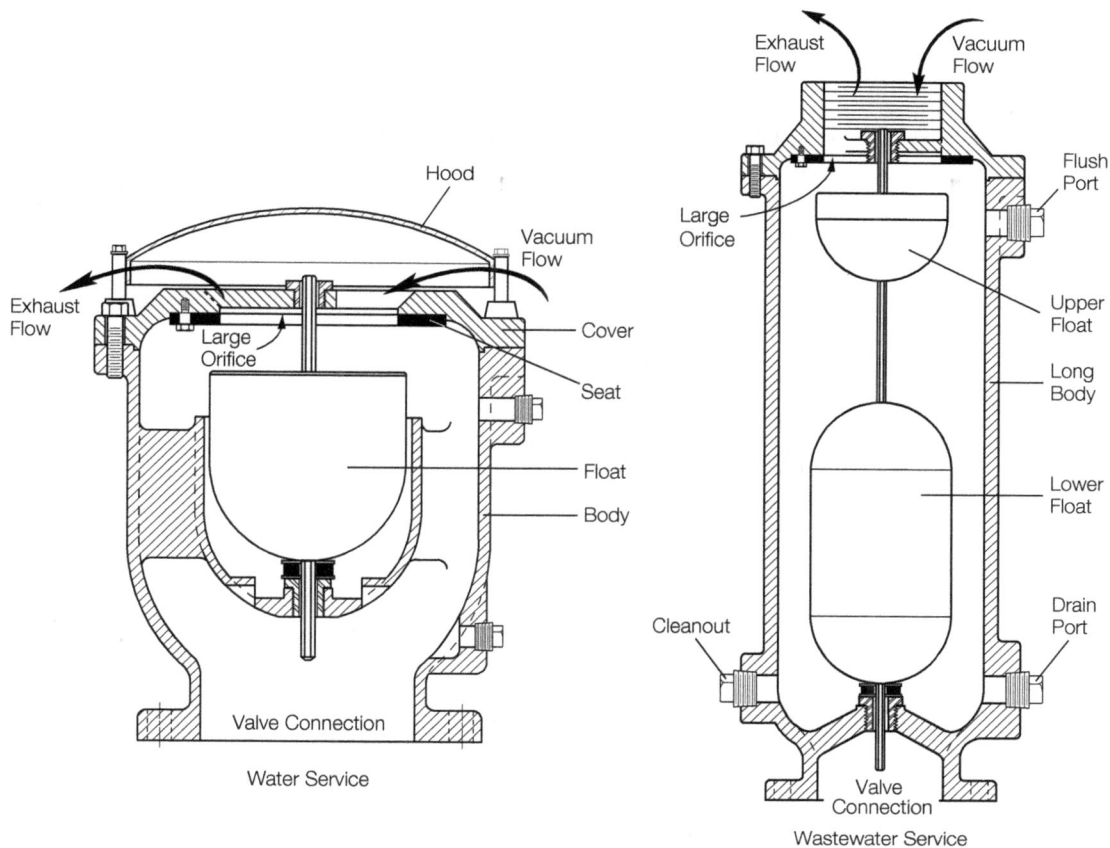

Figure 2-2   Air/vacuum valves

Air/vacuum valves covered by ANSI/AWWA C512 are characterized by orifices having an area equivalent to a circle with a diameter between 1 in. (13 mm) and 20 in. (500 mm) with equal or larger inlet connections. A vacuum breaker is a type of air/vacuum valve that only allows inflow as illustrated in Figure 4-3. As a liquid piping system fills with liquid, the air and wastewater gases in the liquid piping system must be expelled smoothly and uniformly to minimize pressure surges. Likewise, after a pump shutdown or as a liquid piping system drains, air must be admitted to the liquid piping system to prevent the formation of a vacuum that may collapse thin-walled pipes or to prevent the creation of transient pressures as the vapor cavity collapses. At locations of liquid column separation, allowing air to enter the liquid piping system quickly and at sufficient quantities through efficient and properly sized air/vacuum valves or vacuum breakers will help prevent hazardous vapor cavities that may upon collapse cause extreme upsurges or pressure transients.

The operation of an air/vacuum valve is similar to the float operation of the air-release valve except that the orifice diameter is considerably larger and will not open under pressure. An air/vacuum valve is normally open and designed to vent large quantities of air through the orifice. In a float-operated air/vacuum valve, as liquid enters the valve during filling of the system, the float will rise, closing the orifice. Air/vacuum valves once closed WILL NOT REOPEN TO VENT AIR AND WASTEWATER GASES while the liquid piping system is operating under pressure exceeding atmospheric pressure.

COMBINATION AIR VALVES

Combination air valves are designed to perform the same function as air/vacuum valves but, in addition, they will perform the same function as air-release valves and automatically release small pockets of air and wastewater gases from the liquid piping system where the system pressure is greater than atmospheric. Combination air valves can be supplied in a single-body configuration as shown in Figure 2-3 or in a dual-body configuration as shown in Figures 2-4 and 2-5.

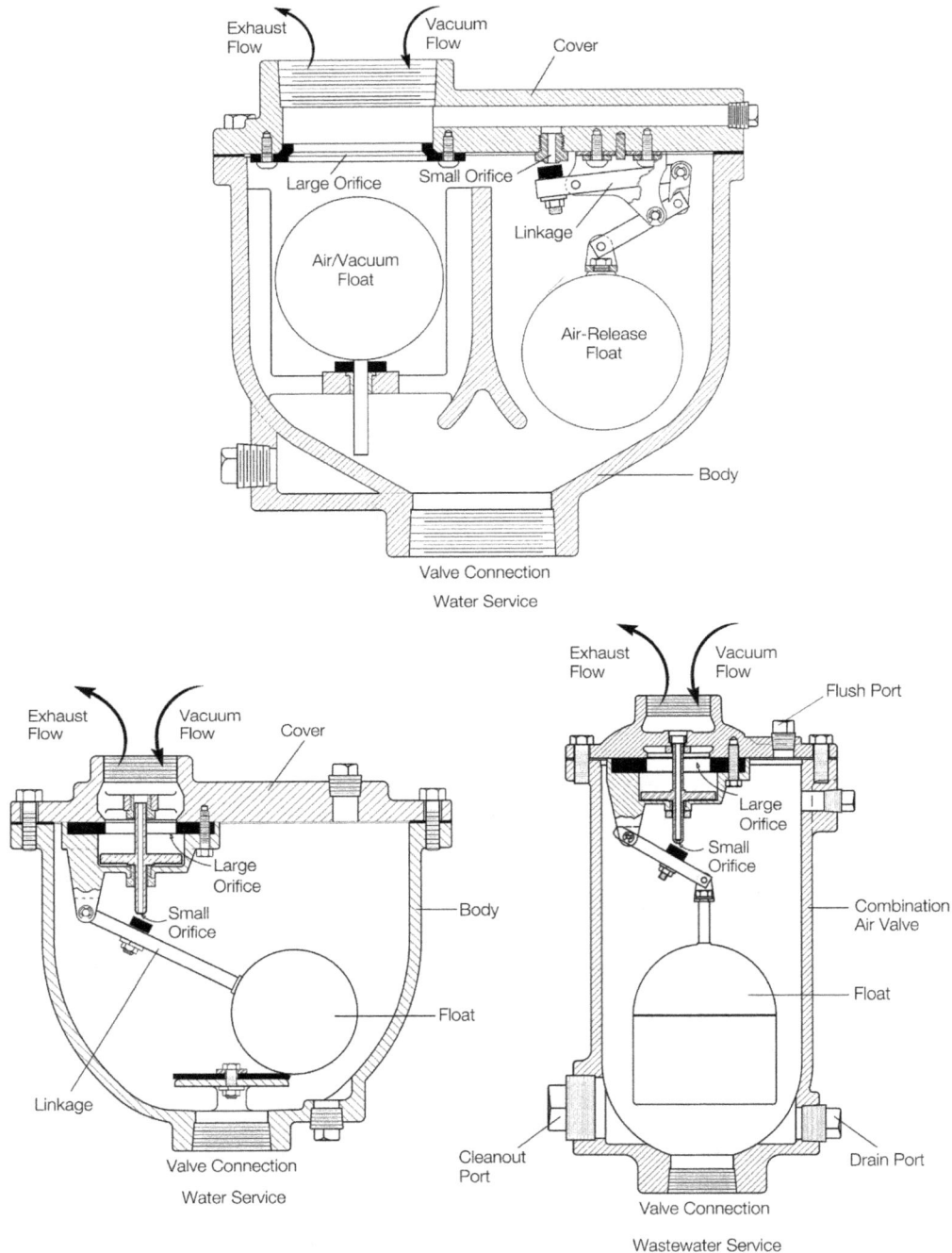

Figure 2-3   Single-body combination air valves

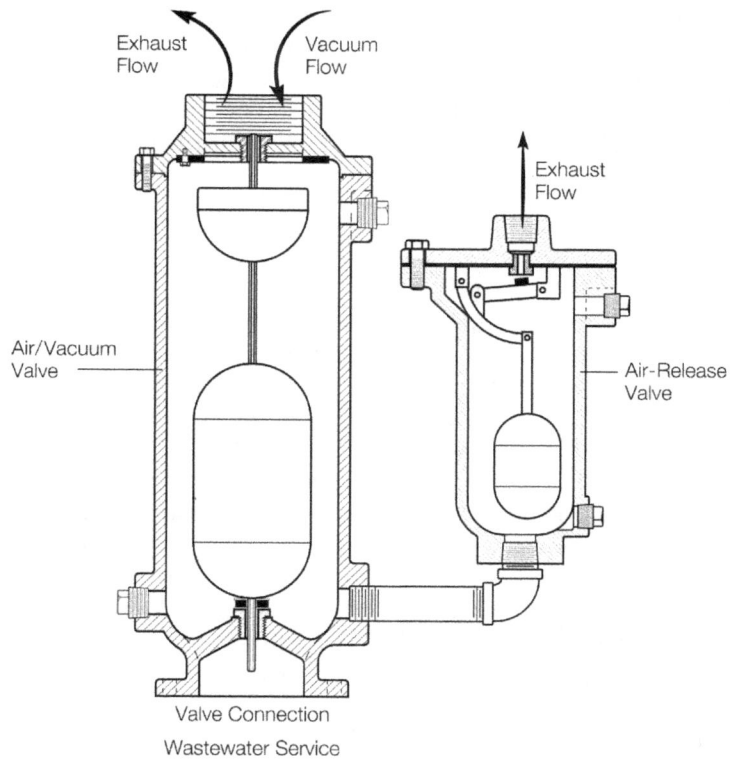

Figure 2-4   Dual-body combination air valves

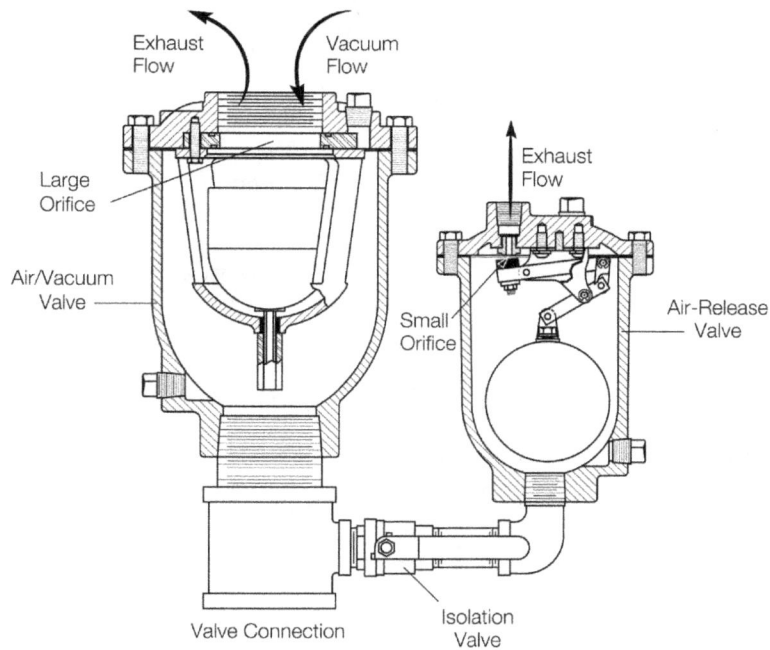

Figure 2-5   Dual-body combination air valve (3 in. [80 mm] and smaller) for water service, bottom-mounted

Wastewater valves can be equipped with backwash appurtenances, which may include an isolation valve, water hose, and quick-disconnect couplings as shown in Figure 2-6. Backwashing is typically a three-step process: (1) Close the isolation valve under the air valve, (2) open the air valve flush port slowly to vent the internal pressure, and (3) flush the air valve with clean water. A cleanout is typically provided when large debris is anticipated as illustrated in Figure 2-3.

AIR VALVE OPTIONAL DEVICES

Air valves are often equipped with optional equipment including slow-closing devices and throttling devices, which control the exhaust rate through the valve.

A slow-closing device, as shown in Figure 2-7, is an adjustable mechanical device to restrict the flow. It is mounted on the inlet or outlet of an air/vacuum valve or combination air valve for water service or on the outlet of an air/vacuum valve or combination air valve for wastewater service. The device contains a disc or closure member with a small opening or multiple ports so that when it closes due to a high exhaust flow rate, the flow area is greatly reduced. Designs vary widely, but the principle of operation is the same. Slow-closing devices are typically used where column separation or vacuum conditions may occur. They allow large volumes of air to rapidly enter the liquid piping system during vacuum conditions, but help to regulate the exhaust flow rate of air and wastewater gases when the liquid columns rejoin to reduce the potential for rapid air valve closure and upsurges or water hammer.

A throttling device is an adjustable mechanical device mounted on the outlet of an air/vacuum valve or combination air valve to control the air and wastewater gases exhaust flow rate out of the valve. As shown in Figure 2-8, the exhaust air and wastewater gases

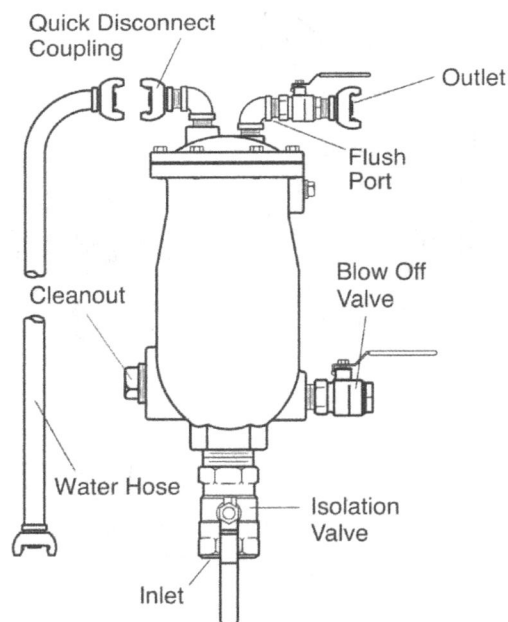

Figure 2-6   Backwash connections

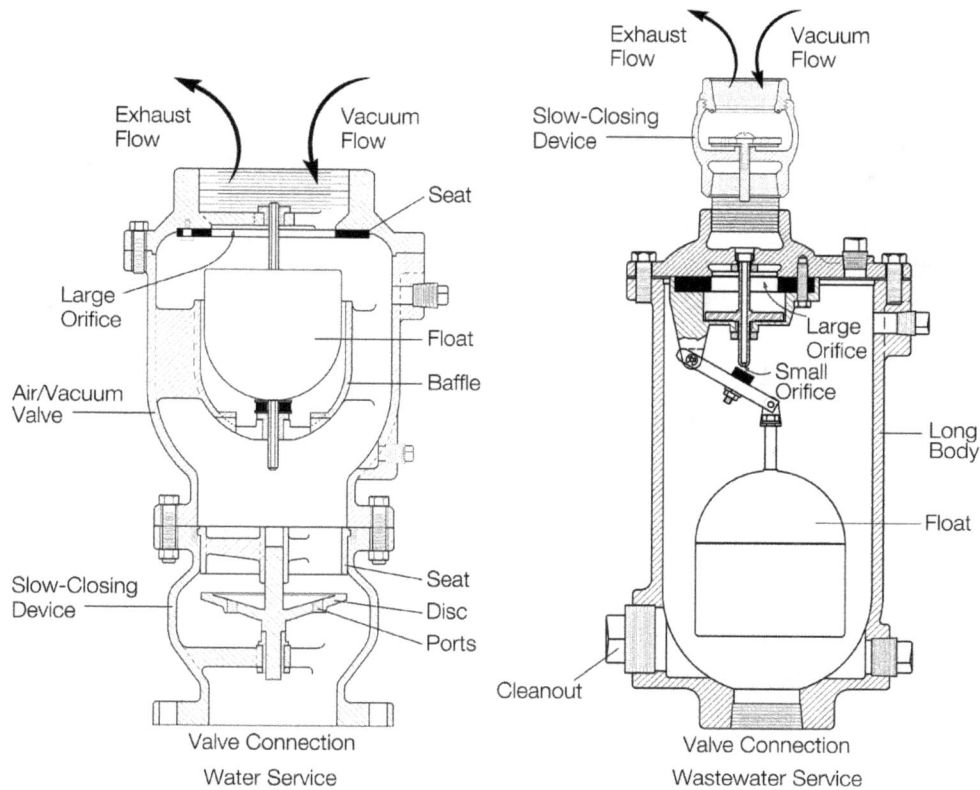

Figure 2-7   Air/vacuum and combination air valves with slow-closing device

enter at the bottom of the device and exit out the side or top. The exhaust rate is controlled by an adjustable disc, which reduces the flow area. These devices can typically restrict the flow area from 5 percent to 100 percent open. The throttling device provides for vacuum flow that is not affected or restricted when the exhaust flow rate is controlled because the discs are spring loaded and are allowed to move open during vacuum flow conditions.

Throttling devices are regularly used on the outlet of air/vacuum or combination air valves as shown in Figure 2-9. When installed on the discharge of well and vertical turbine pumps, they control the rise of the liquid column in the pump and prevent the liquid column from slamming into the downstream check valve (Figure 2-10). Any air and wastewater gases that travel past the check valve typically collect in the pump discharge header and are exhausted through a combination air valve mounted on a riser on the discharge header (Figure 2-11).

Throttling devices are also sometimes used in place of a slow-closing device to restrict air and wastewater gas flow out of the air valve and prevent pressure surges or valve damage. Air valves can be equipped with a vacuum check device on the outlet to prevent the entry of air into the liquid piping system. It may be necessary to prevent the inflow of air in special applications such as a siphon as shown in Figure 3-2. A vacuum check is generally a common soft-seated valve mounted on the air valve to only allow flow out of the piping system.

Vacuum check devices should be used with caution on wastewater valves because they may prevent the draining and flushing of the valves during normal operation.

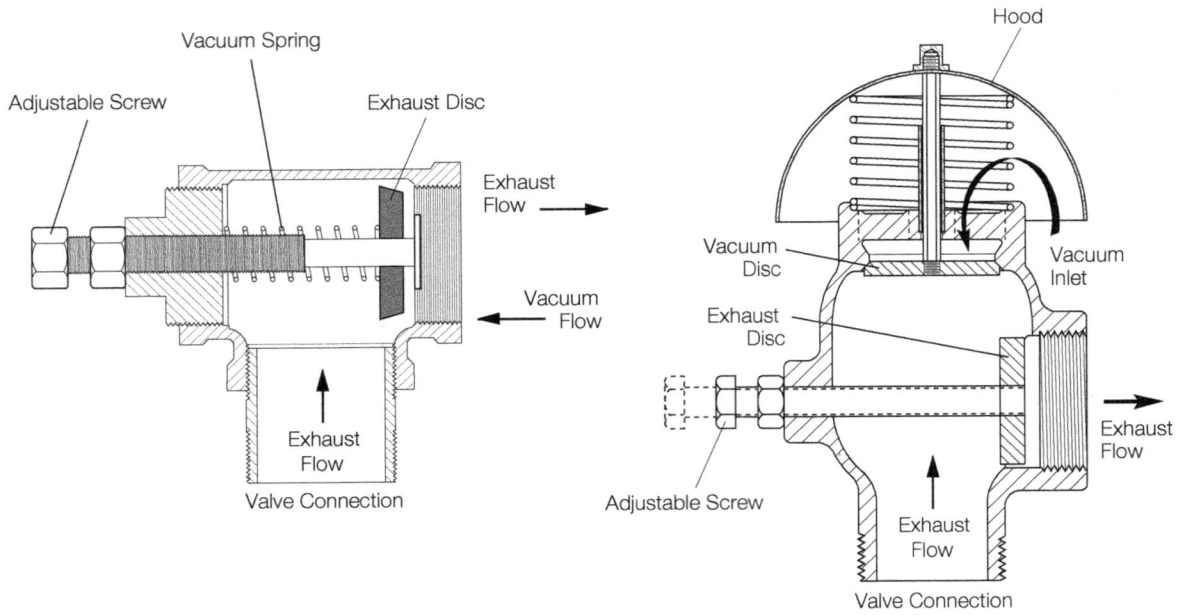

Figure 2-8   Two examples of throttling devices

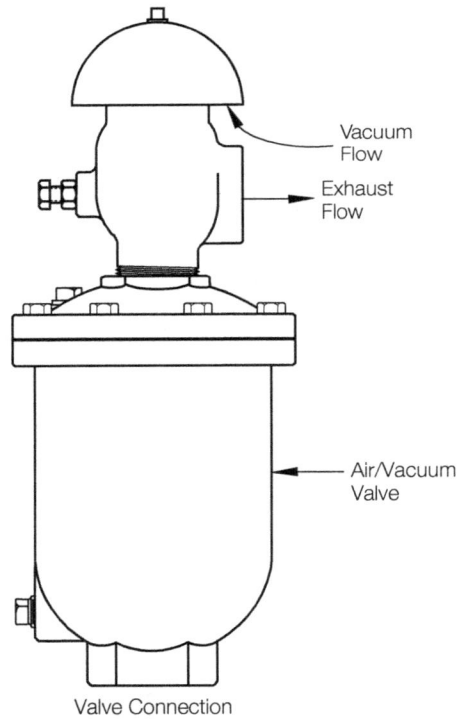

Figure 2-9   Installation of a throttling device

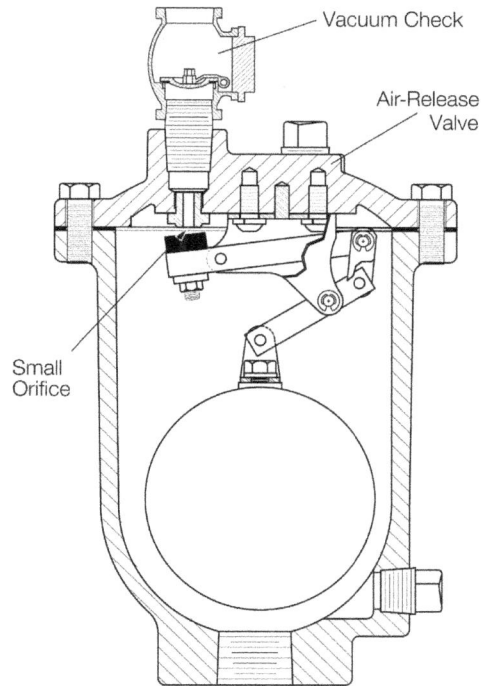

Figure 2-10   Air valve with vacuum check device on the outlet

Figure 2-11   A single-body combination air valve mounted on a pipe header

This page intentionally blank.

Chapter **3**

Locating Air Valves Along a Liquid Piping System at Filling, Steady State Flow, and Drainage

This chapter addresses the location of air valves along a liquid piping system for the elimination of pockets of air and wastewater gases, which could potentially cause binding, and for liquid piping system drainage. The information in this chapter is intended to apply generally to liquid piping systems but may also apply to other situations. This manual does not address the location or use of air valves for downsurge and column separation control, which should be considered for some systems where applicable.

LIQUID PIPING SYSTEM LOCATIONS

The proper location of air-release, air/vacuum, and combination air valves is as important as the proper sizing of the air valve. An improper location can render the valve ineffective. The following guidelines are recommended for the general location and corresponding types of air valves. However, there may be other locations where valves may be deemed necessary based on the liquid piping system design and operation.

Air valves are typically used in liquid piping systems where water or wastewater is being transported to a treatment plant, potable water or reclaimed wastewater is being transported to the end-user, or in other similar applications.

Some relatively small, horizontally level distribution systems with regular service connections and water towers operate with air valves only at the well or high service pumps. However, air valves may be considered on the downstream side of isolation valves. When a segment of the distribution system requires repair, the segment can be filled with water and the air vented through a hydrant, temporary valve, or service connection before putting the system back into service.

When the liquid piping system contains significant high points where binding can occur or when vacuum protection is required, air valves should then be considered to reduce head loss by eliminating entrained air. Also, in these types of distribution systems not having elevated tanks, air valves may be needed to prevent vacuum conditions.

A sample liquid piping system profile illustrating typical valve locations is shown in Figure 3-1 in feet. The horizontal axis is the running length of the liquid piping system, which in the United States is usually expressed using station points. Station points are typically expressed in hundreds of feet, such as 145+32, which is equivalent to 14,532 ft, or in thousands of meters, such as 1+453.00, which is equivalent to 1,453 m. The vertical axis is the elevation of the profile stations relative to a specified horizontal datum. The scale of the vertical axis is typically enlarged to enhance the high and low points of the piping system profile. As indicated in Figure 3-1, the flow is from left to right toward the reservoir. If the flow was in the opposite direction, upslopes become downslopes and the types of air valves may change as described in the following the figure.

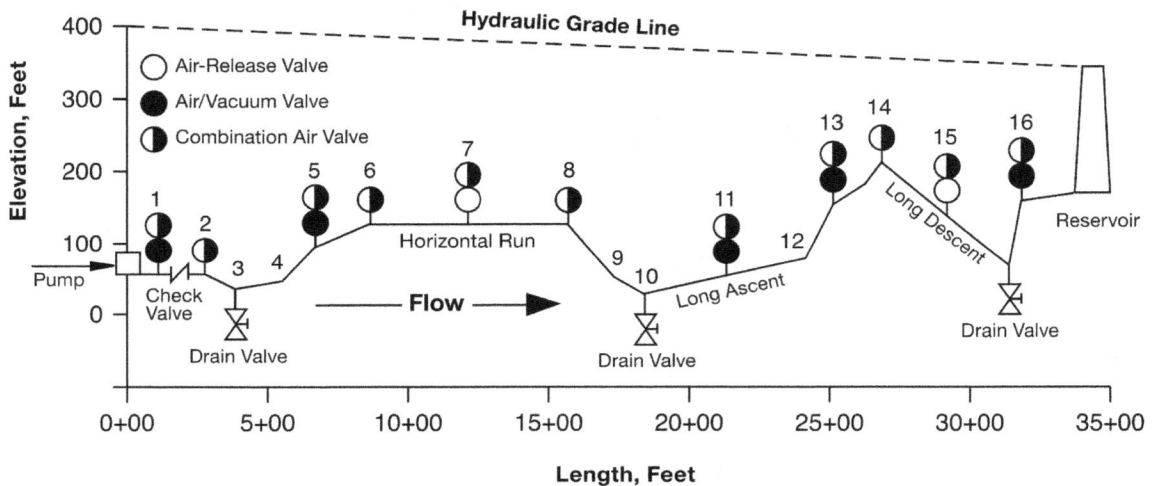

No.	Description	Recommended Types	No.	Description	Recommended Types
1	Pump discharge	Air/Vac or combination	9	Decrease downslope	No valve required
2	Increase downslope	Combination	10	Low point	No valve required
3	Low point	No valve required	11	Long ascent	Air/Vac or combination
4	Increase upslope	No valve required	12	Increase upslope	No valve required
5	Decrease upslope	Air/Vac or combination	13	Decrease upslope	Air/Vac or combination
6	Begin horizontal	Combination	14	High point	Combination
7	Horizontal	Air-Release or combination	15	Long descent	Air-Release or combination
8	End horizontal	Combination	16	Decrease upslope	Air/Vac or combination

Figure 3-1   Sample liquid piping system profile illustrating typical air valve locations

SUGGESTED LOCATIONS AND TYPE

Air valves should be installed at the following locations:

- *High points.* Combination air valves should be considered at liquid piping system high points to provide venting while the liquid piping system is being filled, during normal operation of the liquid piping system, and for air inflow to provide vacuum protection while the pipe is draining. A high point is defined by the higher end of any pipe segment that slopes upward toward the hydraulic gradient or runs horizontal to it, followed immediately by a downsloping segment.

- *Mainline valves* (not illustrated in Figure 3-1). Air/vacuum valves or combination air valves should be considered on the downstream side of mainline valves to facilitate draining of the liquid piping system or a segment of the system from its isolating valve to the draining valve. An air-release valve should be considered on the pressurized side of a mainline valve to facilitate venting of air during the initial filling or testing of a liquid pipeline segment.

- *Increased downslope.* A combination air valve should be considered at abrupt increases in downslope.

- *Decreased upslope.* An air/vacuum valve or a combination air valve should be considered at abrupt decreases in upslope.

- *Long ascents.* An air/vacuum valve or combination air valve should be considered at intervals of ¼ mi (400 m) to ½ mi (800 m) along ascending sections of liquid piping systems.

- *Long descents.* An air-release valve or combination air valve should be considered at intervals of ¼ mi (400 m) to ½ mi (800 m) along descending sections of liquid piping systems.

- *Horizontal runs.* Combination air valves should be considered at the beginning and end of long horizontal sections, and air-release valves or combination air valves should be considered at intervals of ¼ mi (400 m) to ½ mi (800 m) along horizontal sections of pipe. It is difficult to evacuate air and wastewater gases from a long horizontal liquid piping system at low-flow velocities.

- *Transient locations* (not illustrated in Figure 3-1). Other locations may be identified by a transient analysis mathematical model of the distribution system such as identifying the anticipated location of potential column separations. Air valves used to mitigate the effects of column separation should be equipped with slow-closing devices, or vacuum breakers equipped with air-release valves should be considered (Figure 4-3).

- *Flowmeters* (not illustrated in Figure 3-1). Air release valves should be considered upstream of flowmeters to mitigate measurement inaccuracies caused by entrapped air.

- *Deep-well and vertical turbine pumps.* Air/vacuum or combination air valves should be considered on the discharge side of deep-well and vertical turbine pumps to exhaust the air and wastewater gases in the pump column during pump start-up and to allow air back into the line after pump shutdown. Air valves mounted on these types of pumps may require slow-closing devices or throttling devices owing to the violent changes in flow rate during pump cycling. Air-release valves can also be considered for use with time-delayed, power-actuated check valves to release the air and wastewater gases in the pump column slowly under full pump pressure.

- *Siphons* (illustrated in Figure 3-2). To build and maintain a siphon on a section of pipe that extends above the hydraulic gradient and that constantly runs under negative pressure, a combination air valve should be considered on the high point of the siphon to vent the air at pipe filling and during system operation. However, the combination air valve must be equipped with a vacuum check device on the outlet to prevent admitting air into the piping system. When reverse flow is undesirable after pump stoppage, a specialized air/vacuum siphon valve should be considered. A siphon make-and-break valve is designed to vent air during pump start-up, close tight during flowing conditions, and open to break the siphon during reverse-flow conditions by the use of a flow paddle (Figure 3-3). When the siphon is filled by the pump, the main disc lifts to release the air in the pipe. The forward flow pushes against the paddle and holds the side check valve closed. When the pump is stopped and the flow reverses, the paddle opens the side check valve to break the siphon.

- *Penstocks* (see Figure 3-4). A penstock is a relatively short, steep, aboveground pipeline that delivers water from a high-elevation lake or reservoir down to a lower-elevation hydraulic-driven turbine. The penstock operates at the head pressure of the elevated water level and is usually constructed of steel pipe subject to collapse from internal vacuum pressures. If the penstock is equipped with a stop valve on the upstream end, air valves or vacuum breakers should be considered to admit large volumes of air preventing a vacuum condition in the penstock due to gravity flow. As shown in Figure 4-3, a vacuum breaker is a type of air/vacuum valve in that it opens rapidly to allow air flow into the piping system but is normally closed and does not allow air flow out of the system. Vacuum breakers may be float operated like an air/vacuum valve or spring operated.

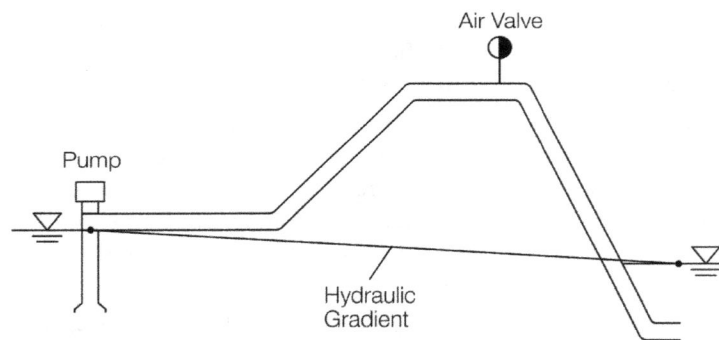

Figure 3-2   Siphon with an air valve

Air
Flow
Out

Air
Flow
In

Flow
Paddle

Pump

Flow

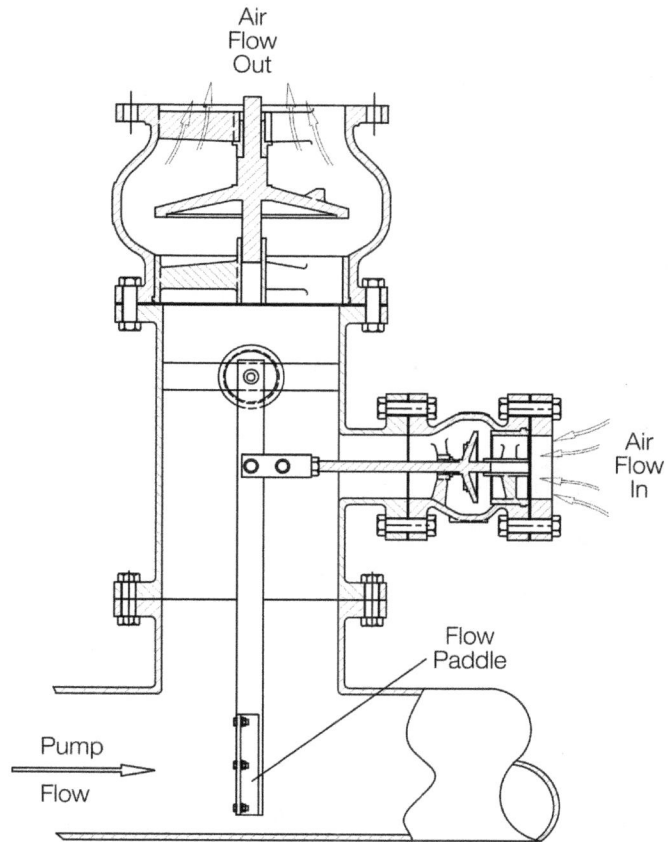

Figure 3-3   Siphon make-and-break air valve with flow paddle

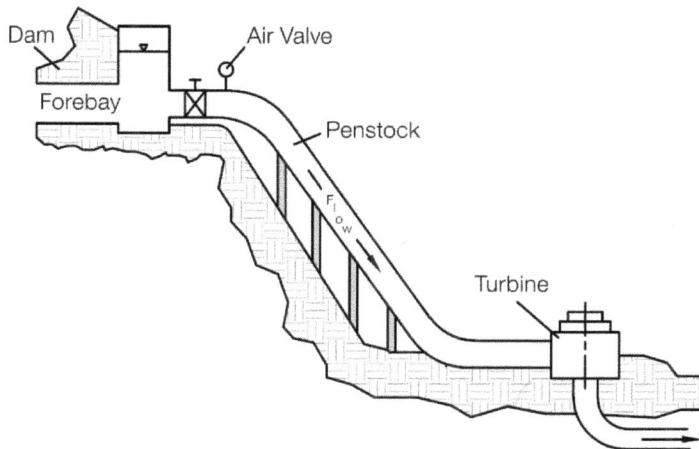

Dam

Air Valve

Forebay

Penstock

Flow

Turbine

Figure 3-4   Penstock

This page intentionally blank.

Chapter **4**

Design of
Valve Orifice Size

It is important to select the proper size valve orifice for the specific location along the liquid piping system. This chapter provides common methodologies used in the water and wastewater industry based on formulas and data tables. Numeric examples are provided for clarity. For specific sizing of valves, refer to manufacturers' charts, graphs, formulas, and software; the figures presented in this chapter only demonstrate the methods used.

SIZING FOR RELEASING AIR AND WASTEWATER GASES UNDER PRESSURE

The orifice size for releasing air and wastewater gases under pressure is generally between ¹⁄₁₆ in. (1.6 mm) and 1 in. (25 mm) in diameter or equivalent area of other shapes; however, the size of the air-release valve inlet connection can range from ½ in. (13 mm) to 6 in. (150 mm) in diameter with the smaller orifices found in the smaller-sized inlet port and higher-pressure valves.

There is no definitive method for determining the amount of air and wastewater gases that may need to be vented from a given liquid piping system. This is because of the difficulty in predicting the quantity of air and wastewater gases that will enter the system or come out of solution as the pressure varies along the liquid piping system. One common method is to provide sufficient capacity to release 2 percent of the liquid flow in terms of air at standard conditions of 60°F (15.6°C) and 14.7 psia (101.4 kPa absolute or 1.0 atm) (Lescovich 1972). This method is based on the 2 percent solubility of air in water at standard conditions and uses orifice sizing data based on air volume at standard conditions (SCFM for free air). Another method is based on the same solubility, but uses the compressed volume of air for orifice sizing (McPherson 2009) and uses orifice sizing data based on air volume at actual conditions (ACFM for pressurized air). Both methods provide similar results. Finally, it should be noted that for wastewater applications where wastewater gases may be generated by the fluid, a higher solubility constant (i.e., 2 to 5 percent) may be considered.

The air and wastewater gases are vented through the orifice of the air-release valve or combination air valve at the liquid piping system working pressure for each valve location. Since high pressures are involved, the applicable flow equation for air flow through an orifice is based on compressible flow. Sonic flow will occur when discharging air and wastewater gases at a pressure exceeding 1.9 times the outlet pressure. Assuming that the outlet pressure is atmospheric (14.7 psia [101 kPa (absolute)]), then any inlet pressure exceeding 1.9 times 14.7 psia (i.e., 28 psia [193 kPa (absolute)] or 27.9 (14.7 × 1.9) psig (90 kPa gauge) will produce sonic flow (Bean 1971). At sonic flow, the air and wastewater gases velocity is limited to the speed of sound, thereby causing a restriction to the air and wastewater gases discharge at higher pressures.

For the purposes of generating Table 4-1, sonic flow and an ideal discharge coefficient of 0.6 were assumed. A discharge coefficient of 0.6 is for a square-edged orifice. Since air and wastewater gases pass through the air valve body and not just through the orifice, there could be additional losses so the actual discharge coefficient of the valve and piping might be different and may vary greatly between different valve designs. The range commonly used is from 0.5 to 0.7 depending on orifice shape and geometry. Air valve capacities at different system pressures should be determined by actual testing for each individual air valve since the discharge coefficient changes from valve to valve and even for the same valve at different pressures. Manufacturers develop coefficients from actual flow tests and typically present flow data in their tables and/or graphs. Therefore, the capacity charts of valve suppliers should be consulted before selecting the final valve size.

The working pressure of an air-release valve is calculated with reference to the maximum hydraulic grade line at the valve and not the pump discharge head. The working differential pressure at the air-release valve location is the difference between the valve elevation and the maximum hydraulic grade line elevation at the valve.

The following example is used to illustrate the approximate orifice size required in an air-release valve. It is important to verify with the supplier that the valve will properly operate with the determined orifice size at the expected maximum line pressure or less. Valve capacity information is presented in tabular form to assist the user. A flow formula is also provided to calculate the capacity of varying orifice diameters at the maximum opening pressure.

ORIFICE SIZING METHOD FOR RELEASING AIR AND WASTEWATER GASES

The orifice sizing method for releasing air and wastewater gases consists of the following four steps:

- **Step 1.** Divide the system flow rate in gallons per minute (gpm) by 7.48 to obtain flow in cubic feet per minute (cfm). If there are multiple valves, the cfm can be divided by the number of air-release valves in the system.

- **Step 2.** Multiply the flow in cfm from step 1 by 0.02 (or 0.02 to 0.05 for gas-containing wastewater) to determine the required venting volumetric rate, expressed in standard cubic feet per minute (SCFM). *Standard* refers to air at the conditions of 60°F (15.6°C) and 14.7 psia (101.4 kPa [absolute]).

- **Step 3.** Determine the working pressure at the valve by subtracting the valve elevation from the maximum hydraulic grade elevation. Express the pressure in pounds per square inch (psi). If the elevations are in feet, multiply by 0.433 to obtain psi.

- **Step 4.** Refer to Table 4-1 and select the orifice diameter that provides the required capacity from step 2 at the pressure from step 3. Consult the available orifice sizes from suppliers and select the valve that meets both the capacity and pressure requirements of the application.

Table 4-1   Air and wastewater gases capacity table (SCFM [L/sec]) of air-release valve orifices ($C_d = 0.6$)

Pressure	Orifice Diameter, in. (mm)									
psi	1/16	3/32	1/8	3/16	1/4	5/16	3/8	7/16	1/2	1
(kPa)	(1.6)	(2.4)	(3.2)	(4.8)	(6.4)	(7.9)	(9.5)	(11.1)	(12.7)	(25.4)
25	1.3	3.0	5.4	12.1	21.5	33.7	48.5	66.0	86.2	344.7
(172)	(.6)	(1.4)	(2.5)	(5.7)	(10.2)	(15.9)	(22.9)	(31.1)	(40.7)	(162.7)
50	2.2	4.9	8.8	19.8	35.1	54.9	79.0	107.5	140.5	561.8
(345)	(1.0)	(2.3)	(4.1)	(9.3)	(16.6)	(25.9)	(37.3)	(50.7)	(66.3)	(265.1)
75	3.0	6.8	12.2	27.4	48.7	76.1	109.5	149.1	194.7	779
(517)	(1.4)	(3.2)	(5.7)	(12.9)	(23.0)	(35.9)	(51.7)	(70.4)	(91.9)	(367.6)
100	3.9	8.8	15.6	35.0	62.2	97.3	140.1	190.6	249.0	996
(689)	(1.8)	(4.1)	(7.3)	(16.5)	(29.4)	(45.9)	(66.1)	(90.0)	(117.5)	(470.0)
125	4.7	10.7	19.0	42.6	75.8	118.5	170.6	232.2	303.3	1,213
(862)	(2.2)	(5.0)	(8.9)	(20.1)	(35.8)	(55.9)	(80.5)	(109.6)	(143.1)	(572.4)
150	5.6	12.6	22.3	50.3	89.4	139.7	201.1	273.7	357.5	1,430
(1,034)	(2.6)	(5.9)	(10.5)	(23.7)	(42.2)	(65.9)	(94.9)	(129.2)	(168.7)	(674.8)
175	6.4	14.5	25.7	57.9	103.0	160.9	231.6	315.3	411.8	1,647
(1,207)	(3.0)	(6.8)	(12.1)	(27.3)	(48.6)	(75.9)	(109.3)	(148.8)	(194.3)	(777.2)
200	7.3	16.4	29.1	65.5	116.5	182.1	262.2	356.8	466.1	1,864
(1,379)	(3.4)	(7.7)	(13.7)	(30.9)	(55.0)	(85.9)	(123.7)	(168.4)	(219.9)	(879.6)
225	8.1	18.3	32.5	73.2	130.1	203.3	292.7	398.4	520.3	2,081
(1,551)	(3.8)	(8.6)	(15.3)	(34.5)	(61.4)	(95.9)	(138.1)	(188.0)	(245.6)	(982)
250	9.0	20.2	35.9	80.8	143.7	224.5	323.2	439.9	574.6	2,298
(1,724)	(4.2)	(9.5)	(16.9)	(38.1)	(67.8)	(105.9)	(152.5)	(207.6)	(271.2)	(1,085)
275	9.8	22.1	39.3	88.4	157.2	245.7	353.8	481.5	628.9	25,515
(1,896)	(4.6)	(10.4)	(18.5)	(41.7)	(74.2)	(115.9)	(166.9)	(227.2)	(296.8)	(1,187)
300	10.7	24.0	42.7	96.1	170.8	266.9	384.3	523.0	683.2	2,732
(2,068)	(5.0)	(11.3)	(20.1)	(45.3)	(80.6)	(125.9)	(181.3)	(246.8)	(322.4)	(1,289)

Example

A liquid piping system with a flow rate of 10,500 gpm requires an air-release valve at a location with a valve elevation of 600 ft and a hydraulic grade line elevation of 831 ft.

1. 10,500 gpm/7.48 = 1,404 cfm
2. 1,404 × 0.02 = 28 SCFM
3. (831 − 600) × 0.433 = 100 psi
4. Select 3/16-in. (4.8-mm) orifice from Table 4-1 that provides 35.0 SCFM (16.5 L/sec) at 100 psi (689 kPa)

The capacity information shown in Table 4-1 is based on the compressible flow equation and sonic flow (Crane [1982] *Technical Paper No. 410*).

$$Q = 678\, Yd^2 C_d \sqrt{\Delta P P_1/(TS_g)} \qquad \text{(Eq. 4-1)}$$

Where:

Q = flow rate, SCFM

Y = expansion factor, 0.71 for air flow (Crane [1982], *Technical Paper No. 410*)

d = orifice diameter, in.

C_d = coefficient of discharge, 0.6

ΔP = differential pressure, 0.47 P_1 (for sonic flow)

P_1 = inlet pressure, psia (liquid piping system pressure + 14.7 psi) (assumes sea level atmospheric pressure of 14.7 psia; pressure will vary with altitude)

T = inlet temperature, 520 degrees Rankine

S_g = specific gravity, 1.0 (for air) and 0.7 to 1.0 (for wastewater gases)

For subsonic conditions where system pressures are generally less than 13 psig (90 kPa [gauge]):

$$Q = 14.77\, d^2 [\Delta P (P + 14.7)]^{1/2} \qquad \text{(Eq. 4-2)}$$

Where:

P = system pressure, psig

SIZING FOR LIQUID PIPING SYSTEM FILLING

For the initial filling of a liquid piping system, air should be vented at the same volumetric rate as the system is being filled. In many cases, only one pump is operated until the line is full. The recommended procedure, however, is to use a portable pump or a throttled flow rate to fill the liquid piping system at a gradual rate to prevent surges in the line. As a general rule, the suggested filling velocity based on the full pipe cross section should not exceed 1 ft/sec (0.3 m/sec) because when the flow reaches the far end of the system and suddenly stops, it may cause a transient pressure increase of about 50 psi (345 kPa) for every 1 ft/sec (0.3 m/sec) of velocity. For more information, see the discussion on water hammer in chapter 5.

The volumetric rate of air from initial filling is vented to atmosphere at a typical differential pressure of 2 psi (13.8 kPa). The following method may be used to approximate the orifice size required for filling. Generic tables and formulas are provided to suit the preference of the user.

The applicable flow equation is based on compressible flow through a short nozzle or tube where there is no heat transfer to the air. Also, it is assumed that the valve is at sea level and at a temperature of 60°F (15.6°C). At high altitudes or extreme temperatures, equations that take these factors into consideration should be used. For the purpose of generating Table 4-2, a discharge coefficient of 0.6 is used. A discharge coefficient of 0.6 is an approximation for a square-edged orifice. Air valve capacities at different system pressures should be determined by actual testing for each individual air valve since the discharge coefficient changes from valve to valve, and even for the same valve at different pressures. Therefore, valve suppliers' capacity charts should be consulted before selecting the final valve size.

Table 4-2   Air and wastewater gas discharge table (SCFM [L/sec]) of large orifices (C_d = 0.6, T = 60°F [15.6°C], and sea level)

Differential Pressure	Orifice Diameter, in. (mm)											
psi	1	2	3	4	6	8	10	12	14	16	18	20
(kPa)	(25.4)	(50.8)	(76.2)	(101.6)	(152)	(203)	(254)	(305)	(355)	(406)	(457)	(508)
1.0	68	271	611	1,086	2,443	4,343	6,786	9,772	13,300	17,372	21,986	27,143
(6.9)	(32)	(128)	(288)	(512)	(1,153)	(2,049)	(3,202)	(4,611)	(6,276)	(8,198)	(10,375)	(12,809)
1.5	83	330	743	1,321	2,973	5,285	8,257	11,891	16,185	21,139	26,754	33,030
(10.3)	(39)	(156)	(351)	(623)	(1,403)	(2,494)	(3,897)	(5,611)	(7,638)	(9,976)	(12,625)	(15,587)
2.0	95	379	853	1,516	3,411	6,064	9,475	13,644	18,571	24,255	30,698	37,899
(13.8)	(45)	(179)	(402)	(715)	(1,610)	(2,862)	(4,471)	(6,438)	(8,763)	(11,446)	(14,486)	(17,885)
2.5	105	421	946	1,683	3,786	6,731	10,517	15,144	20,612	26,922	34,074	42,066
(17.2)	(50)	(199)	(447)	(794)	(1,787)	(3,176)	(4,963)	(7,146)	(9,727)	(12,705)	(16,079)	(19,851)
3.0	114	457	1,028	1,828	4,114	7,313	11,427	16,454	22,396	29,252	37,022	45,706
(20.7)	(54)	(216)	(485)	(863)	(1,941)	(3,451)	(5,392)	(7,765)	(10,569)	(13,804)	(17,471)	(21,569)
3.5	122	489	1,099	1,955	4,398	7,818	12,216	17,591	23,944	31,273	39,581	48,865
(24.1)	(58)	(231)	(519)	(922)	(2,075)	(3,689)	(5,765)	(8,301)	(11,299)	(14,758)	(18,678)	(23,059)
4.0	129	517	1,164	2,069	4,655	8,275	12,929	18,618	25,341	33,099	41,891	51,717
(27.6)	(61)	(244)	(549)	(976)	(2,196)	(3,905)	(6,101)	(8,786)	(11,959)	(15,619)	(19,768)	(24,405)
4.5	136	543	1,221	2,170	4,883	8,681	13,564	19,532	26,586	34,724	43,948	54,256
(31.0)	(64)	(256)	(576)	(1,024)	(2,304)	(4,097)	(6,401)	(9,217)	(12,546)	(16,386)	(20,739)	(25,604)
5.0	141	565	1,272	2,261	5,086	9,042	14,129	20,345	27,692	36,169	45,777	56,515
(34.5)	(67)	(267)	(600)	(1,067)	(2,400)	(4,267)	(6,667)	(9,601)	(13,068)	(17,068)	(21,602)	(26,669)

Orifice Sizing Method for Liquid Piping System Filling

Assume air valve is at sea level and 60°F [15.6°C].

- **Step 1.** Calculate the venting flow rate in SCFM using:

$$Q = q \ (0.134 \ \text{ft}^3/\text{gal}) \ \frac{(\Delta P + 14.7 \ \text{psi})}{(14.7 \ \text{psi})} \qquad \text{(Eq. 4-3)}$$

Where:

Q = flow rate, SCFM
q = fill rate, gpm
ΔP = differential pressure, 2 psi

- **Step 2.** Refer to Table 4-2 and select the orifice diameter that provides the required flow at the selected venting pressure.

Example

A 66-in. pipeline will be filled at a flow rate of 10,500 gpm (1 ft/sec) and the air valve will vent the air at a pressure of 2 psi.

1. $Q = (10,500) \ (0.134) \ [(2.0 + 14.7) / 14.7] = 1,598$ SCFM
2. Referring to Table 4-2, at 2 psi, select a 4-in. orifice that will vent 1,780 SCFM.

SIZING FOR LIQUID PIPING SYSTEM DRAINING

When it becomes necessary to drain a liquid piping system, the system should be drained at a controlled rate of approximately 1 to 2 ft/sec (0.3 to 0.6 m/sec) to minimize pressure transients. An air valve at the high point adjacent to the draining location must be properly sized to admit air at the same volumetric rate as the system is being drained.

SIZING FOR GRAVITY FLOW

A power failure or line break may result in a sudden change in the flow velocity, potentially causing column separation from the resulting gravity flow. The gravity flow may create excessive vacuum conditions at the adjacent high points. Most small-size and medium-size pipes commonly used in the water industry can withstand a complete vacuum, but there could be serious damage to fittings, accessories, joints, seals, and so on at a vacuum of only –5 psig (–34 kPa[gauge]), which can cause leakage and/or contaminant intrusion (Gullick et al. 2004). However, owing to lower stiffness ratios, large-diameter pipes may collapse from negative internal pressures. Therefore, sizing air valves for gravity flow conditions is important to maintaining the integrity of the liquid piping system. Air valves at high points should be sized to allow the inflow of air to minimize negative pressures in the pipe and prevent possible damage to pump seals, equipment, or the pipe itself.

When sizing an air valve orifice for gravity flow, the pipe slope will determine the volume of air required to prevent excessive vacuum. Appropriate air valves should be provided at all high points with the orifice sized to allow the required inflow of air to replace the liquid in the pipe. Figure 4-1 illustrates the required inflow of air required for various pipe sizes and slopes.

Pipe Diameter, in.

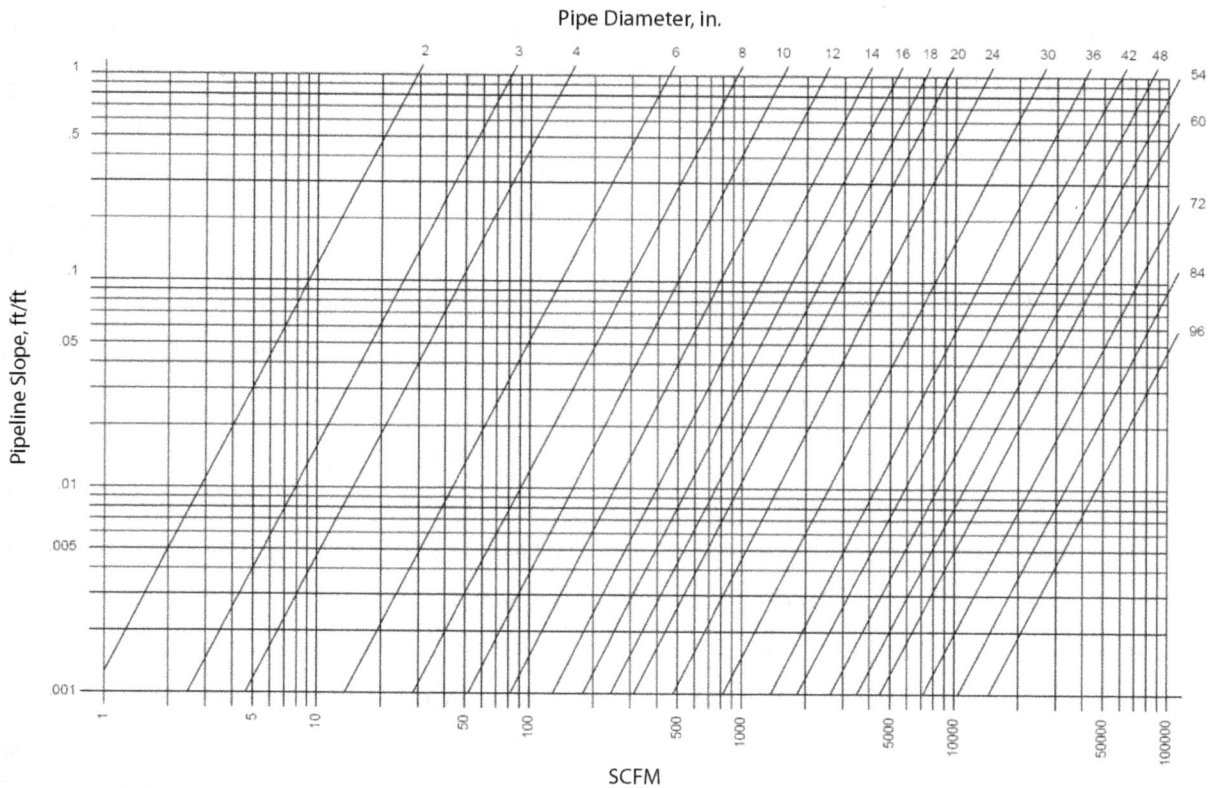

Figure 4-1   Inflow of air for gravity flow, SCFM

The orifice sizing of an air valve for inflow is typically based on the lower of –5 psig (–34 kPa[gauge]) or the allowable negative pressure below atmospheric pressure for the pipe with a suitable safety factor such as 4:1. Sonic flow will occur when the outlet-to-inlet pressure ratio (Bean 1971) falls below 0.53. Knowing that the inlet pressure is atmospheric (14.7 psia [101 kPa (absolute)]), then any negative system pressure below 7.8 psia (54 kPa [absolute]) or –7 psig (–48 kPa [gauge]) (i.e., vacuum) will produce sonic flow. Because the flow will be sonic and restricted, flow volume will not increase beyond –7 psi (–48 kPa) differential.

If gravity flow occurs in a pipe with a change in slope where the pipe lower section has a steeper slope than the upper section, then an air/vacuum valve should be considered at the location where the pipe changes slope. The gravity flow will be greater in the pipe section with the steeper slope. The air/vacuum valve orifice should be sized so that the inflow of air at this point equals the difference in the two flow rates at the allowable negative pressure.

The applicable flow equation is based on compressible flow through a short nozzle or tube where there is no heat transfer to the air and subsonic flow. For the purposes of estimating circular orifice sizes, a discharge coefficient, C_d, of 0.6 was used to generate Table 4-3. A discharge coefficient of 0.6 is an approximation of a square-edged circular orifice. Therefore, valve suppliers' capacity charts should be consulted before selecting the final valve size. Further, as in any piping system where multiple components are piped together, the resultant flow capacity should be calculated with consideration of all of the devices and piping in the system.

Table 4-3   Air inflow table of large orifices (SCFM [L/sec]) (C_d = 0.6)

Differential Pressure	Orifice Diameter, in. (mm)											
psi	1	2	3	4	6	8	10	12	14	16	18	20
(kPa)	(25.4)	(50.8)	(76.2)	(101.6)	(152)	(203)	(254)	(305)	(355)	(406)	(457)	(508)
1.0	66	264	593	1,055	2,374	4,220	6,593	9,495	12,923	16,879	21,363	26,374
(6.9)	(31)	(124)	(280)	(498)	(1,120)	(1,991)	(3,111)	(4,481)	(6,098)	(7,965)	(10,081)	(12,446)
1.5	79	317	713	1,268	2,853	5,072	7,925	11,411	15,532	20,287	25,676	31,698
(10.3)	(37)	(150)	(337)	(598)	(1,346)	(2,393)	(3,740)	(5,385)	(7,330)	(9,573)	(12,116)	(14,958)
2.0	90	359	808	1,436	3,231	5,745	8,976	12,926	17,594	22,980	29,083	35,905
(13.8)	(42)	(169)	(381)	(678)	(1,525)	(2,711)	(4,236)	(6,100)	(8,302)	(10,844)	(13,724)	(16,944)
2.5	98	394	886	1,575	3,543	6,298	9,841	14,171	19,289	25,194	31,886	39,365
(17.2)	(46)	(186)	(418)	(743)	(1,672)	(2,972)	(4,644)	(6,687)	(9,102)	(11,889)	(15,047)	(18,576)
3.0	106	423	951	1,691	3,804	6,763	10,567	15,217	20,712	27,052	34,238	42,269
(20.7)	(50)	(199)	(449)	(798)	(1,795)	(3,191)	(4,987)	(7,181)	(9,774)	(12,766)	(16,157)	(19,947)
3.5	112	447	1,007	1,789	4,026	7,158	11,184	16,104	21,920	28,630	36,235	44,735
(24.1)	(53)	(211)	(475)	(844)	(1,900)	(3,378)	(5,278)	(7,600)	(10,344)	(13,511)	(17,099)	(21,110)
4.0	117	468	1,054	1,874	4,215	7,494	11,710	16,862	22,951	29,977	37,939	46,838
(27.6)	(55)	(221)	(497)	(884)	(1,989)	(3,536)	(5,526)	(7,957)	(10,830)	(14,146)	(17,903)	(22,103)
4.5	122	486	1,094	1,945	4,377	7,782	12,159	17,509	23,831	31,126	39,394	48,635
(31)	(57)	(230)	(516)	(918)	(2,066)	(3,672)	(5,738)	(8,262)	(11,246)	(14,689)	(18,590)	(22,951)
5.0	125	502	1,129	2,007	4,515	8,026	12,541	18,059	24,581	32,105	40,633	50,165
(34.5)	(59)	(237)	(533)	(947)	(2,131)	(3,788)	(5,918)	(8,522)	(11,600)	(15,150)	(19,175)	(23,673)

Orifice Sizing Method for Gravity Flow

- **Step 1.** Determine the allowable negative pressure for the pipe with consideration for a reasonable safety factor. Consult the pipe manufacturer for the maximum recommended negative pressure. For low-stiffness, large-diameter steel pipe, the collapse pressure can be estimated by the general formula for collapse of thin-walled steel cylinders (AWWA Manual M11 [2004]). The formula is applicable to a pipe submerged or in an aboveground environment. Pipes used in buried service with good soil compaction are not prone to vacuum collapse (Watkins and Tupac 1994).

$$P_c = 66,000,000 \, (t/d)^3 \qquad \text{(Eq. 4-4)}$$

Where:

P_c = collapsing pressure, psi
t = pipe wall thickness, in.
d = mean diameter of pipe, in.

The allowable differential pressure for sizing is then found by the formula

$$\Delta P = P_c / SF \qquad \text{(Eq. 4-5)}$$

Where:

ΔP = differential pressure, psi
SF = safety factor, dimensionless

The choice of safety factor (i.e., 3.0 or 4.0) is at the discretion of the liquid piping system designer. When the pipe is not subject to collapse, a differential pressure of 5.0 psi (34 kPa) is commonly used.

- **Step 2.** Calculate the slope of the pipe (S) as the change in elevation divided by horizontal distance (i.e., rise over run expressed in the same units, ft/ft).

- **Step 3.** Determine the required air inflow in SCFM from Figure 4-1 by matching the pipe slope against the pipe diameter. For increases in downgrade and decreases in upgrade, compute the difference between the flows in the lower and upper sections of pipe. Flow rates can also be calculated using common flow formulas such as Hazen-Williams, Manning, and Darcy-Weisbach or the following formula:

$$Q = 0.0472 \, C \sqrt{S \, ID^5} \qquad \text{(Eq. 4-6)}$$

Where:

Q = flow rate, SCFM

C = Chezy coefficient, 110 (iron), 120 (concrete), 130 (steel), 190 (polyvinyl chloride). NOTE: The coefficient, C, varies for different pipe roughness and is different from the C-factor associated with the Hazen-Williams formula.

S = pipe slope, ft/ft
NOTE: The S in the actual Chezy, Hazen-Williams, and Darcy-Weisbach equations actually refers to the slope of the hydraulic grade line and not the pipe slope. For steeper pipe slopes, the difference between the hydraulic grade line slope and the pipe slope can be significant. This is one reason the maximum slope in Figure 4-1 only reaches 1 ft/ft (45°).

ID = pipe inside diameter, in.

- **Step 4.** Refer to Table 4-3 for selecting the orifice diameter that provides the required flow in SCFM at the permissible differential pressure.

Example

Using the aboveground 24-in.-ID by ⅛-in.-thick steel pipeline and Figure 4-2, calculate the minimum orifice diameter at Stations 10+00 (assuming a line break at Station 0+00), 25+00 (assuming a line break at Station 20+00), and 40+00 (assuming a line break at Station 20+00).

1. d = $ID + t$ = 24.000 in. + 0.125 in. = 24.125 in.
   P_c = 66,000,000 (0.125 in. / 24.125 in.)^3 = 9.2 psi (From Eq. 4-4)
   Assuming a safety factor of 4.0,
   ΔP = 9.2 psi / 4.0 = 2.3 psi (From Eq. 4-5)

2. S_1 = 40 ft / 1,000 ft = 0.04
   S_2 = 40 ft / 500 ft = 0.08
   S_3 = 20 ft / 1,500 ft = 0.013

3. For S_1 = 0.04 and ID = 24, Figure 4-1 provides Q_1 = 3,000 SCFM
   For S_2 = 0.08 and ID = 24, Figure 4-1 provides Q_2 = 4,050 SCFM
   For S_3 = 0.013 and ID = 24, Figure 4-1 provides Q_3 = 1,900 SCFM

To size the orifice at Station 25+00, Q_{25+00} = 4,050 – 1,900 = 2,150 SCFM.

4. For Station 10+00, use Table 4-3 and select a 6-in. orifice with an inflow capacity of approximately 3,400 SCFM at 2.3 psi that exceeds Q_1 of 3,000 SCFM.

   For Station 25+00, use Table 4-3 and select a 6-in. orifice with an inflow capacity of approximately 3,400 SCFM at 2.3 psi that exceeds Q_2 of 2,150 SCFM.

   For Station 40+00, use Table 4-3 and select a 6-in. orifice with an inflow capacity of approximately 3,400 SCFM at 2.3 psi that exceeds Q_3 of 1,900 SCFM.

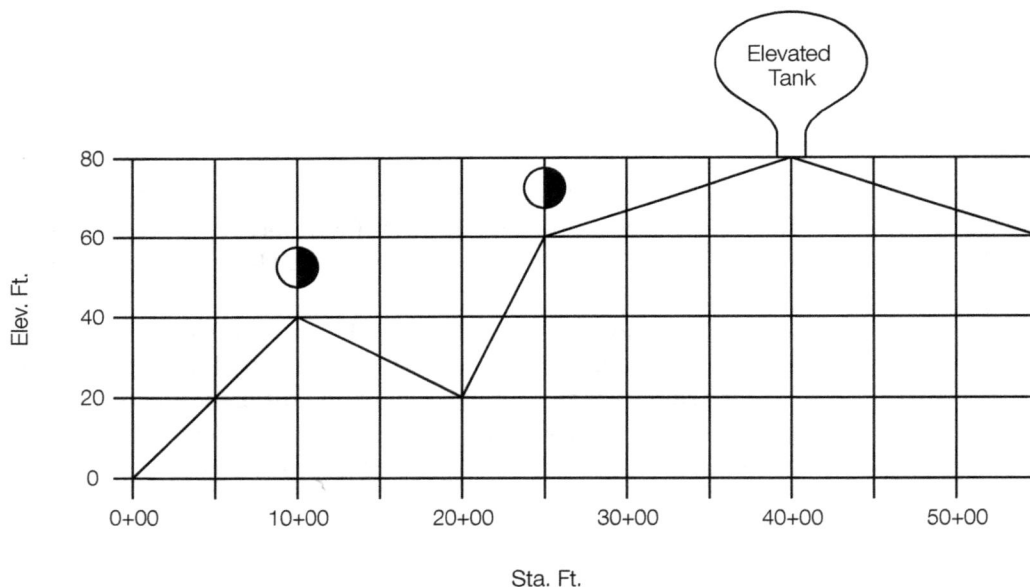

Figure 4-2   Example of air valve locations for gravity flow along a pipe

ORIFICE SIZING FOR PARTIAL RUPTURE

In the previous section, the gravity flow sizing assumed that there was a complete rupture of the pipe diameter and full gravity flow. Full gravity flow represents a worst-case scenario. In this sizing analysis, the pipe system designer selects the maximum rupture size for which to provide vacuum protection. The flow rate is based on free discharge from a rupture with a calculated area using the following equations.

$$Q_r = C_d A \sqrt{2g\Delta h} \qquad \text{(Eq. 4-7)}$$

Where:

Q_r = required air flow rate for rupture protection, ft³/sec
C_d = orifice coefficient (commonly 0.6)
A = area of rupture, ft²
g = gravitational acceleration (32.2 ft/sec²)
Δh = elevation distance between the air valve location and the rupture, ft

If the rupture is assumed to be circular:

$$Q_r = \frac{C_d \pi D^2}{4} \sqrt{2g\Delta h} \qquad \text{(Eq. 4-8)}$$

Where:

D = equivalent diameter of the rupture, ft

If the rupture is more than 30 percent the size of the pipe:

$$Q_r = F_C \frac{C_d \pi D^2}{4} \sqrt{2g\Delta h} \qquad \text{(Eq. 4-9)}$$

Where:

F_C = correction factor

and:

$$F_C = \sqrt{\frac{1}{1-\left(\frac{D}{D_p}\right)^4}} \qquad \text{(Eq. 4-10)}$$

Where:

D_p = pipe diameter (OD), ft

These equations do not include head losses in the pipe. The Hazen-Williams formula is used to determine head losses, using the flow rate, Q_r, calculated in Eq. 4-9 above:

$$h_f = L_a \left(\frac{4.73 Q_r^{1.85}}{C_{HW}^{1.85} D_P^{4.87}} \right) \qquad \text{(Eq. 4-11)}$$

Where:

> h_f = head loss, ft
> L_a = actual pipe segment length, ft
> C_{HW} = Hazen-Williams coefficient for the pipe

and:

$$L_a = \sqrt{\Delta h^2 + L^2} \qquad \text{(Eq. 4-12)}$$

Where:

L = length of the horizontal projection of the pipe segment (distance in the pipe profile), ft

Using Eq. 4-9 and replacing Δh with $(\Delta h - h_f)$ yields the following:

$$Q_r = F_C \frac{C_d \pi D^2}{4} \sqrt{2g(\Delta h - h_f)} \qquad \text{(Eq. 4-13)}$$

SIZING FOR VACUUM CONDITIONS

There are special situations requiring the application of air valves, such as the control of liquid column separation and the minimizing of subsequent pressure transients. Sizing of these valves is usually included in the transient analysis of a liquid piping system using a computer program and is beyond the scope of this manual.

This manual of practice assumes a line break or partial rupture of the pipe at a lower elevation, such as Station 20+00 in Figure 4-2, resulting in gravity flow, which may cause a vacuum condition or column separation at the higher elevations, i.e., Stations 10+00, 25+00, and 40+00.

In some cases, such as large-diameter pipes subject to collapse, the size of the air valve calculated in the preceding sections, "Orifice Sizing Method for Gravity Flow" and "Orifice Sizing for Partial Rupture," may be beyond the size range of readily manufactured valves. In these cases, it is suggested to install clusters of valves. Another alternative is to use a vacuum breaker available in larger sizes in combination with an air valve to provide the needed inflow capacity, as shown in Figure 4-3.

The sizing of air valves for vertical turbine deep-well pump discharge service is highly dependent on the specific characteristics of the air valve and sometimes the pump. Therefore, these applications should be based on the published sizing recommendations of the air valve supplier. Deep-well pump applications are described further in chapter 5.

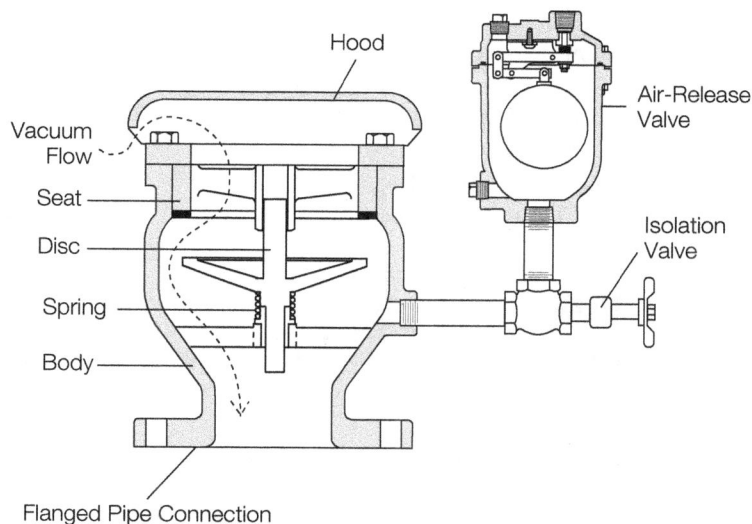

Figure 4-3   Vacuum breaker with air-release valve

AIR-RELEASE VALVE SELECTION

The following information is recommended for selecting the correct air-release valve for venting accumulated air and wastewater gases during liquid piping system operation:

- Compliance with ANSI/AWWA C512, latest edition
- Orifice size from the section "Sizing for Releasing Air and Wastewater Gases Under Pressure"
- NPT or flanged inlet size
- Design pressure of each valve
- Minimum and maximum operating pressure of each valve
- Valve construction materials including special corrosion-resistant materials for wastewater service
- Type of installation (in-plant, in-vault, or outdoor)

The selection of a larger orifice or inlet size is generally acceptable as long as the maximum operating pressure is not exceeded.

For a given orifice size (e.g., ⅛ in. [3 mm]), several inlet sizes may be available (e.g., ½ in. [13 mm] to 6 in. [150 mm]). The inlet size should be as large as possible to maximize the air/water exchange in the pipe connection. Also, the pipe connection should never be less than the inlet size of the air-release valve. For wastewater, the minimum inlet size should be at least 2 in. (50 mm), and 3 in. (80 mm) is preferable to avoid clogging.

The maximum design pressure of an air-release valve is related to the construction of the valve body and the mechanical advantage of the float leverage mechanism. The valve must have sufficient mechanical advantage to allow the weight of the float to pull the seal away from the orifice. Valves with large orifices (i.e., greater than ⅛ in. [3 mm]) or high operating pressures (i.e., greater than 175 psi [1,206 kPa]) will usually employ a compound lever mechanism with a series of levers and pivot pins. It is important for the valve to have a maximum working pressure greater than the highest expected operating pressure at the specific valve location. Air valves for water and wastewater services are tested at minimum pressures of 20 psig (138 kPa). If air valves are required to operate at a lower pressure, the lower pressure should be specified. Wastewater air valves may need to operate and seal at pressure below 20 psig (138 kPa) during gravity flow and static non-flow conditions.

Typical options for air-release valves include special corrosion-resistant construction or a vacuum check on the valve outlet to prevent air from reentering the system during negative pressure conditions (Figure 2-10, page 15). For wastewater, all internal parts including pins, levers, and internal surfaces should be corrosion resistant.

AIR/VACUUM VALVE SELECTION

The following information is recommended for selecting the correct air/vacuum valve for venting air during liquid piping system filling and admitting air during negative pressure conditions:

- Compliance to ANSI/AWWA C512, latest edition
- Orifice size
- Inlet size and type of connection
- Design pressure of each valve
- Minimum and maximum operating pressures of each valve
- Valve construction materials including special corrosion-resistant materials for wastewater service
- Type of installation (in-plant, in-vault, or outdoor)
- Type of outlet connection (threaded, flanged, or hooded)

The orifice size must be sufficient to meet all of the requirements for:

1. Venting air during filling per section "Sizing for Liquid Piping System Filling"
2. Admitting air during draining per section "Sizing for Liquid Piping System Draining"
3. Admitting air during line break per section "Sizing for Gravity Flow"
4. Admitting air at liquid column separation

Select a valve size that satisfies all four requirements. In accordance with ANSI/AWWA C512, the inlet size for an air/vacuum valve generally matches the orifice size. For wastewater air valves, the larger the inlet connection, the greater the valve's ability to capture air and wastewater gas bubbles flowing along the top of the pipe. Larger inlet connections may also be considered for wastewater applications to prevent clogging.

When large-size air/vacuum valves are required, it may be possible to use multiple air/vacuum valves in a cluster or in nearby locations on the pipeline. Multiple air/vacuum valves provide redundancy, which can continue to provide vacuum protection when one valve is being serviced or becomes inoperable. Moreover, smaller valves are readily available and may shorten the duration of installation.

When multiple valves are used, it is important to understand that the area of an orifice is related to the square of the diameter of the valve size. Equivalent valve sizes and orifice areas are shown in Table 4-4 for reference. For example, a cluster of two 4-in. valve orifice area (eq. 25.1 in.²) does not equal a single 8-in. valve (eq. 50 in.²); it would require four 4-in. (eq. 50 in.²) or two 6-in. valves (eq. 56.5 in.²) to equal one 8-in. valve orifice area (eq. 50 in.²). Clusters can also consist of valves of different sizes. In this case, the flow areas from the various size combinations are summed and compared to the required flow area.

An alternate to a valve cluster is the use of a vacuum breaker, which can be provided in very large diameters (Figure 4-3).

Table 4-4 Nominal areas of single and multiple air valves

| Valve Configuration | Nominal Air Valve Flow Area (in.²) | | | | | | | | | | | |
| | Valve Size (in.) | | | | | | | | | | | |
	1	2	3	4	6	8	10	12	14	16	18	20
Single valve	0.79	3.14	7.1	12.6	28.3	50	79	113	154	201	254	314
Two-valve cluster	1.57	6.28	14.1	25.1	56.5	101	157	226	308	402	509	628
Three-valve cluster	2.36	9.42	21.2	37.7	84.8	151	236	339	462	603	763	942

Oversized air/vacuum valves used where the potential for column separation exists or where surges can occur should include slow-closing devices and throttling devices to control the discharge of air and wastewater gases upon column return by slowing down the returning liquid column and thus preventing water hammer and local pressure surges. The maximum pressure rating of the valve will influence the seat material in the valve. Typically, air/vacuum valves rated for high pressure (i.e., greater than 300 psi [2,068 kPa]) and large-diameter valves (i.e., greater than 14 in. [350 mm]) may be equipped with hard nonmetallic seats or stainless-steel seats containing O-ring seals.

In some liquid piping systems, a large orifice is needed to prevent a vacuum at high points, but using the same size orifice for expelling the air when the system pressure recovers may result in a secondary surge when the air is rapidly expelled and returning liquid slams into the closing air valve. In these cases it is recommended to use a slow-closing device or throttling device (Lingireddy et al. 2004).

Typical options for air/vacuum valves include special corrosion-resistant construction, screened hoods, and slow-closing devices or throttling devices mounted on the inlet or outlet to minimize valve-induced pressure surges.

COMBINATION AIR VALVE SELECTION

The following information is recommended for selecting the correct combination air valve for venting air during filling, admitting air during negative pressure conditions, and venting accumulated air and wastewater gases during liquid piping system operation:

- Compliance to ANSI/AWWA C512, latest edition
- Sizes of air-release and air/vacuum orifices
- Inlet size and type of connection
- Design pressure of each valve
- Minimum and maximum operating pressures of each valve
- Valve construction materials including special corrosion-resistant coatings and materials for wastewater service
- Type of installation (in-plant, in-vault, or outdoor)
- Type of outlet connection (threaded, flanged, hooded, or other)
- Body configuration (single body or dual body)

The orifice size must be sufficient to meet all of the requirements for:

1. Venting accumulated air and wastewater gases under pressure per section "Sizing for Releasing Air and Wastewater Gases Under Pressure"
2. Venting air during filling per section "Sizing for Liquid Piping System Filling"

3. Admitting air during draining per section "Sizing for Liquid Piping System Draining"
4. Admitting air during line break per section "Sizing for Gravity Flow"
5. Admitting air at liquid column separation

The selected combination air valve configuration should satisfy all five requirements.

Single-body configurations are typically more economical. They are more compact, are less likely to freeze, and are tamper resistant. Also, if the air-release valve is being serviced, the air/vacuum valve may be kept in service. Single-body configurations are typically limited in size. Dual-body configurations are available in larger sizes and consist of an air-release valve piped to an air/vacuum valve. Many combinations and ranges of capacities are therefore available.

The inlet size for an ANSI/AWWA C512 combination air valve generally matches the orifice size of the air/vacuum valve but may be larger to improve the collection of air. Combination air valves used where the potential for column separation exists or where surges can occur should include slow-closing devices to control air and wastewater gas discharge at column return by slowing down the returning liquid column and thus preventing water hammer and local pressure surges. Another alternative is to use a vacuum breaker available in larger sizes in combination with an air valve to provide the needed inflow capacity. The design pressure of the valve must also include the ability to vent air through the air-release orifice at the expected maximum pressure at that location in the liquid piping system.

For wastewater air valves, the larger the inlet connection, the greater the valve's ability to capture air bubbles flowing along the top of the pipe. Larger inlet connections may also be considered for wastewater applications to prevent clogging.

Typical options for combination air valves include special corrosion-resistant construction and coatings, screened hoods, and slow-closing devices or throttling devices mounted on the inlet or outlet to minimize induced pressure surges.

REFERENCES

ANSI/AWWA C512. Latest edition. Air Release, Air/Vacuum, and Combination Air Valves for Water and Wastewater Service. Denver, CO: AWWA.

AWWA. 2004. Manual M11, *Steel Pipe: A Guide for Design and Installation*. 4th ed. Denver, CO: AWWA.

Bean, H.S. 1971. *Fluid Meters, Their Theory and Application*. 6th ed. New York: ASME.

Crane Co., Engineering Division. 1982. *Flow of Fluids Through Valves, Fittings and Pipe*. Technical Paper No. 410. New York: Crane Co.

Crane Co., Engineering Division. 1982.

Giles, R.V. 1962. *Theory and Problems of Fluid Mechanics and Hydraulics*, 2nd ed. (Schaum's Outlines). New York: McGraw-Hill.

Gullick, R.W., M.W. LeChevallier, R.C. Svindland, and M.J. Friedman. 2004. Occurrence of Transient Low and Negative Pressures in Distribution Systems. *Jour. AWWA*, 96(11):52–66.

Lescovich, J.E. 1972. Locating and Sizing Air-Release Valves. *Jour. AWWA*, 64(7):457–461.

Lingireddy, S., D.J. Wood, and N. Zloczower. 2004. Pressure Surges in Pipeline Systems Resulting From Air Releases. *Jour. AWWA*, 96(7):88–94.

McPherson, D.L. 2009. Air Valve Sizing and Location: A. Prospective. In *Proc., Pipelines 2009: Infrastructures Hidden Assets.* Reston, VA: American Society of Civil Engineers (ASCE).

Tchobanoglous, George, Metcalf and Eddy, 1981. *Wastewater Engineering: Collection and Pumping of Wastewater.* New York: McGraw-Hill.

Watkins, R.K., and G.J. Tupac. 1994. Vacuum Design of Welded Steel Pipe Buried in Poor Soil. In *Proc., Hydraulics of Pipelines.* Reston, VA: ASCE, pp. 226–238.

This page intentionally blank.

Chapter **5**

Water Hammer Effects

Water hammer is a sudden change in pressure resulting from rapid changes in flow velocity within the liquid piping systems and is also referred to as *surge* or *transient pressure* (AWWA Manual M11 [2004]). These sudden changes in pressure are only a phase in an event referred to as a *pressure transient event,* but also known as a water hammer event or a surge event. A pressure transient event includes both upsurges and downsurges that are greatly influenced by each other. This is where air valves play an important role. Water hammer is an extremely complex phenomenon requiring computer analysis; however, the use of general operating principles will minimize its effects. This chapter presents some applications for air valves in systems where water hammer may occur.

AIR/VACUUM AND COMBINATION AIR VALVES

To minimize the effects of water hammer during filling of a liquid piping system, it is recommended that the system filling velocity based on the full pipe cross section be maintained at 1 ft/sec (0.3 m/sec) or less because when the flow reaches the far end of the system and suddenly stops, it may cause a transient pressure increase of about 50 psi (345 kPa) for every 1 ft/sec (0.3 m/sec) of velocity. Properly designed air/vacuum or combination air valves will allow air and wastewater gases to exhaust from the liquid piping system relatively unrestricted. However, when the last of the air and wastewater gases escapes the system, the air/vacuum or combination air valve may shut abruptly in response to the liquid reaching the valve float. The resulting collision between adjacent columns of liquid in the vicinity of the valve may cause a rapid deceleration of the liquid in the pipe, resulting in a surge (Tullis 1989). Air valves may be equipped with slow-closing or flow-throttling devices to minimize the abrupt closing of the air/vacuum or combination air valves.

Air/vacuum or combination air valves are provided on thin-wall piping systems to protect against pipe collapse under negative pressure conditions. These pipes are especially prone to water hammer effects during the filling operations since the orifice diameter required for collapse criteria provides minimal air discharge regulation, especially at excessive filling rates. For these and other installations where large-diameter air valves are used, it is important to provide for strict control of the filling rate. This may require the throttling of the pump discharge flow rate or throttling the gravity supply flow rate during the filling operation. Generally, a filling rate that limits the liquid velocity to 1 ft/sec (0.3 m/sec) is acceptable (Jones et al. 2008). Air/vacuum valves, combination air valves, and vacuum breaker valves can, in many instances, prevent cavitation water hammer by controlling the downsurge, thus preventing the formation of a significant vapor cavity, the collapse of which when the liquid columns return can cause extreme upsurges.

AIR VALVES AT VERTICAL TURBINE OR WELL PUMPS

Air/vacuum or combination air valves installed on the discharge piping of vertical turbine or deep-well pumps are subject to water hammer problems similar to those encountered in the filling of liquid piping systems. Air and wastewater gases needs to be vented from the pump column upon start-up. Otherwise, air and wastewater gases may be delivered into the discharge piping when the check valve opens. Uncontrolled discharge of air and wastewater gases and the abrupt closure of air/vacuum or combination air valves on pump discharge applications can lead to significant pressure transients.

To minimize water hammer effects, the pump discharge flow rate may be controlled at start-up, or either slow-closing devices or air-throttling devices may be incorporated into the air/vacuum valve or combination air valve design. These devices, some of which are specially manufactured for vertical turbine and deep-well pump installations, generally control the exhaust rate and closure speed of the air/vacuum and combination air valves. It is important to note that the slow-closing devices and air-throttling devices are effective in suppressing water hammer only when placed adjacent to the pump between the pump and the check valve or pump control valve. Figure 5-1 shows an example of the proper locations of an air/vacuum valve or combination air valve equipped with either a slow-closing or air-throttling device.

Another important function of air/vacuum or combination air valves and their components is to provide vacuum protection to the pump column, seals, and other parts, at pump shutoff, especially with deep-well pumps. In most cases, a combination air valve is required because dissolved air and wastewater gases are constantly being released out of solution at the pump, and the gases should therefore be exhausted before reaching the check valve to minimize cavitation and pitting, consequently enhancing the liquid piping system.

Air-release valves are often connected directly to the pump column to release air and wastewater gases and protect the mechanical seals. Air-release valves can also be used with time-delayed, power-actuated pump discharge control valves to release air and wastewater gases in the pump column slowly under full pump pressure before the control check valve opens.

AIR VALVES IN LIQUID PIPING SYSTEMS

The presence of air and wastewater gases in a liquid piping system may substantially reduce the conveyance capacity of the pipe. During water hammer conditions, entrapped air and wastewater gases may magnify the pressure transient event. Entrapped air and wastewater gases can store energy and cause check-valve slamming. If pockets of air and

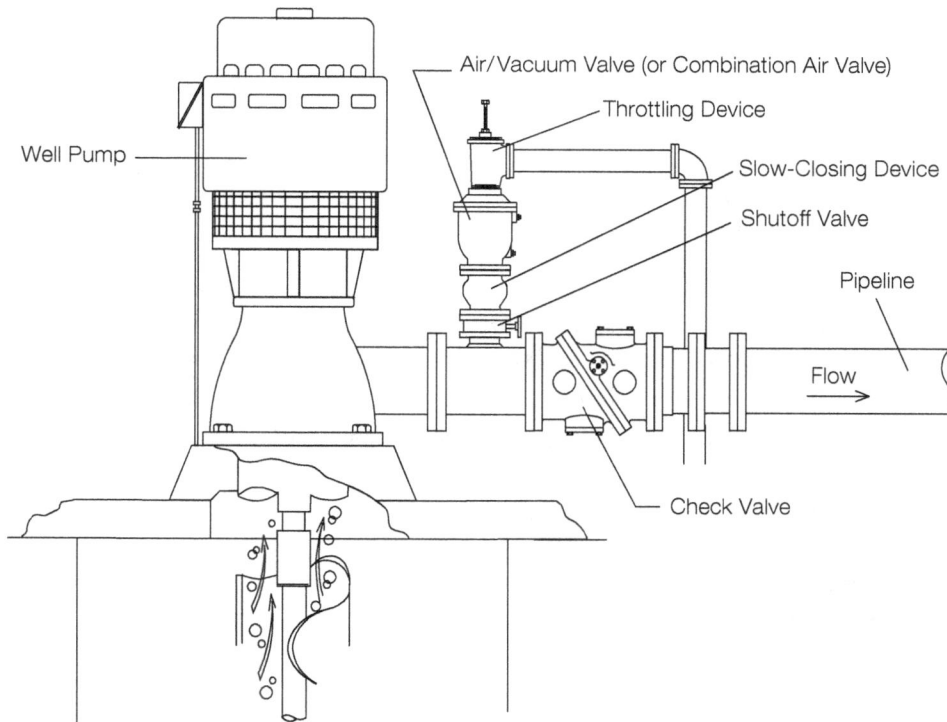

Figure 5-1   Air/vacuum valve or combination air valve at a well pump

wastewater gases become dislodged, water hammer can be initiated by a change in flow velocity when the pockets of air and wastewater gases pass through restrictions, through partially open valves, or from one high point to another.

Conversely, air and wastewater gases can also help mitigate peak transient pressures by acting as an inline accumulator where air and wastewater gases absorb energy by slowing the propagation of the transient pressure wave. However, the use of trapped air and wastewater gases in the system to mitigate transient pressures is not recommended owing to the more significant detrimental effects to the flow capacity of the pipe and the inherent instability of a system with two phases (fluid and gas).

Some general guidelines for minimizing the effects of air and wastewater gases in liquid piping systems are as follows (Tullis 1989):

1. *Fill slowly.* As a general rule, the suggested filling velocity based on the full pipe cross section should not exceed 1 ft/sec (0.3 m/sec) velocity.

2. *Install properly sized air/vacuum valve or combination air valves* so air and wastewater gases are not released at high pressure or high flow rates during pipe filling.

3. *Lay the pipe to a set grade and install air valves at high points.* If the terrain is flat, install air valves at regular intervals.

4. *Flush the system at moderate velocities, 2 to 4 ft/sec (0.6 to 1.2 m/sec), and low pressure* to move the air and wastewater gases to the air valves.

5. *Install air valves upstream of control valves* so air and wastewater gases do not pass through modulating valves.

6. *Use combination air valves* wherever possible to accommodate the flow of air and wastewater gases during filling, draining, and operational accumulation.

7. *Limit the air-and-wastewater-gases discharge rate through air valves* to avoid sonic velocities.

User-friendly software is readily available for locating and sizing air valves from most air valve manufacturers.

Water hammer in liquid piping systems can also be analyzed with special computer programs that mathematically model the transient response of the system during pump starts and stops. Water hammer software can provide immediate feedback of the effects on system performance of suggested air valve locations and sizes including:

- Valve size and location effects during system filling
- Identification of additional (not obvious) locations
- Effectiveness of alternate locations and sizes
- Documentation and consistency of valve locations and sizing

Studies have shown a strong correlation between analysis and system performance (Kroon et al. 1984).

REFERENCES

AWWA. 2004. Manual M11, *Steel Pipe: A Guide for Design and Installation*, 4th ed. Denver, CO: AWWA.

Jones, G.M., R.L. Sanks, G. Tchobanoglous, and B.E. Bosserman II, eds. 2008. *Pumping Station Design*, rev. 3rd ed. Boston: Butterworth-Heinemann.

Kroon, J.R., M.A. Stoner, and W.A. Hunt. 1984. Water Hammer: Causes and Effects. *Jour. AWWA*, 76(11):39–45.

Tullis, J.P. 1989. *Hydraulics of Pipelines: Pumps, Valves, Cavitation, Transients*. New York: John Wiley and Sons.

Chapter **6**

Installation, Operation, Maintenance, and Safety

To ensure that the air valve will operate properly, reasonable care is needed in handling, installation, and maintenance. This chapter provides the basic information for using air valves, but it is important that the information provided with the supplied valve be carefully reviewed and followed.

INSTALLATION

Air Valve Instruction Manual

The air valve instruction manual supplied by the manufacturer should be reviewed in detail before installing an air valve. At the jobsite prior to installation, the air valve should be visually inspected and any packing or foreign material in the interior portion of the valve should be removed. The nameplate information on the air valve should be verified to ensure that the valve coincides with that specified.

Air Valve Location and Connections

The air valve should be installed as close to the pipe as possible. Long interconnecting piping to the air valve should be avoided when possible, and all interconnecting piping must slope upward toward the valve and be sized appropriately to accommodate the required flow of air and wastewater gases. Typically, the interconnecting piping will be greater in diameter than the large orifice of the air valve to facilitate the required flow. The farther the air valve is from the pipe, the greater in diameter the connecting pipe should be.

The air/vacuum orifice is sized to exhaust air during pipeline filling and to provide air intake for vacuum protection. When the interconnecting pipe branch is long, it becomes a dead-ended force main or charged pipeline. Since the air valve orifice size was designed for the main pipe, the greater the difference in size between the main pipe and the interconnecting pipe, the higher the flow velocity in the interconnecting pipe and the greater the resulting surge at valve closure. For this reason, air valves at the end of long interconnecting pipes, especially on force mains that operate in cycles, should always have slow-closing devices.

Small-diameter, long interconnecting pipes hinder air valve vacuum and downsurge protection. When the air valve is close to the main pipe, it reacts very fast to a downsurge caused by drainage and/or liquid column separation, thus mitigating damaging vacuum conditions or water hammer. But if the air valve is at the end of a long, small-diameter interconnecting pipe, liquid in the interconnecting pipe blocks the air entry to the main pipe. Since the diameter of the main pipe is much greater than the diameter of the interconnecting pipe, the drainage flow or the column separation flow is much greater than the return liquid flow in the interconnecting pipe. So there can be a great delay in terms of pressure transients before air entering the air valve reaches the main pipe. For this reason, the longer the interconnecting pipe, the larger its size should be. The air valve manufacturer should be consulted for sizing when long connecting piping is needed.

A shutoff valve of the same diameter as the interconnecting pipe, but no smaller than the air valve inlet connection, should be installed between the air valve and the top of the pipe to facilitate maintenance (Figure 6-1). The shutoff valve should be of the full-port type or low–head-loss type to avoid any obstruction that could possibly affect the capacity of the air valve, and the valve should be located as close to the main pipe as possible.

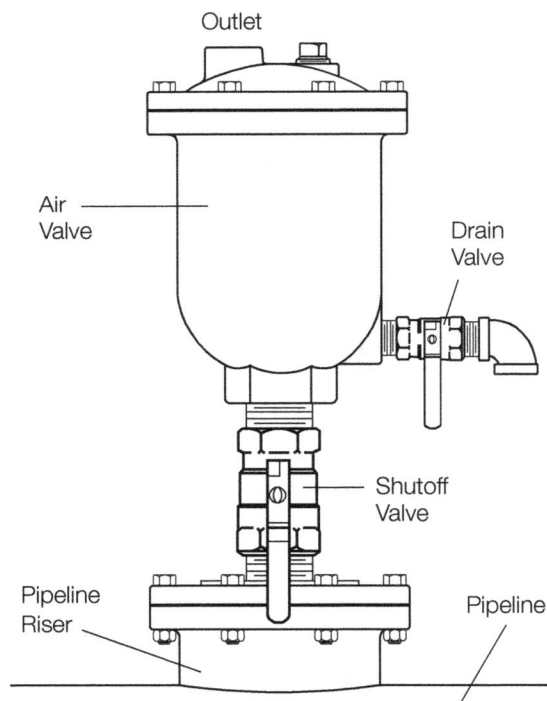

Figure 6-1   Installation of an air valve with shutoff valve

For air/vacuum valves, the size of the connection to the top of the main pipe should equal or exceed that of the air valve large orifice and is dependent on the diameter of the main pipe. For air-release valves and for combination air valves, an air collection trap in the form of an enlarged pipeline riser should be installed on the main pipe leading to the air valve connection (Figure 6-1). This is necessary in order to capture air and wastewater gas bubbles passing the air valve connection.

For liquid piping systems 12 in. (300 mm) and smaller, the diameter and height of the riser should be equal to the pipeline diameter. For pipelines greater than 12 in. (300 mm) and less than 60 in. (1,500 mm), the diameter and height of the riser should be 0.6 times the pipeline diameter. For pipelines greater than 60 in. (1,500 mm), the diameter and height of the riser should be 0.35 times the pipeline diameter (Van Vuuren et al. 2004).

Air Valve Linings and Coating

Internal and external air valve corrosion should be controlled by applying proper lining and coatings where necessary or by making use of corrosion-resistant ANSI/AWWA-C512–compliant metallic materials. This is especially essential in wastewater applications where exposure to heat and corrosive environments can cause rapid corrosion of the air valve. Air valve linings are also important in wastewater applications to facilitate proper valve operation where exposure to solids and greases can cause clogging. All fasteners should be made from corrosion-resistant material (i.e., high-grade stainless steel) or be protected from corrosion by plating or coating. As an alternative to linings and coatings, valves with all-stainless-steel construction or special alloys for corrosive service can be used.

Air Valve Protection From Contamination

Air valves with top-threaded openings should be covered with a protective cap, U-bend, or elbow so that foreign materials cannot externally enter through the top of the valve. Inline reinforced screens are available for use in U-bend–type covers to prevent malicious tampering with the liquid piping system. Air valves can be supplied with large metal hoods equipped with heavy-duty screens that cover the valve discharge opening to prevent the entry of rodents and bird nests.

Noise Control

When discharging large volumes of air and wastewater gases, air valves can create objectionable or hazardous sound levels up to 100 dB, similar to a large air compressor. Air-release valves can emit high-pitched sounds because they exhaust air and wastewater gases that reach sonic flow velocities. Large air/vacuum and combination air valves can exhaust large volumes of air and wastewater gases that at times can sound like a jet engine. Appropriate precautions should be taken such as plumbing the valve discharge to a protected area, installing noise reduction hoods, or installing an appropriately sized silencer. These sounds can be annoying to the general public and can result in complaints; therefore local law enforcement agencies and fire departments should be made aware of outdoor air valve installations that are immediately adjacent to residential, commercial, and other areas accessible by the general public.

Protection of Air Valves From Freezing

If air valves are exposed to freezing conditions, the smaller connecting piping can freeze, trapping liquid in the valve, which then in turn freezes and, because of the expansion of frozen liquid, the frozen liquid can collapse the valve float or crack the valve body.

Air valves can be insulated and equipped with heat tracing to prevent freezing if electrical power is available. In colder climates, air valves are typically installed below-ground within vaults.

Another effective solution for water applications is the use of an automatic freeze protection valve installed in the drain port of the air valve body (Figure 6-2). A freeze protection valve consists of a spring-loaded thermostatic element that senses fluid temperature. When it falls below a specified lower set point temperature (i.e., 35°F [1.7°C]), the valve modulates open, allowing liquid to flow by the sensor. The valve will remain open as long as the temperature of the liquid flowing by the sensor is less than a second specified higher set point temperature (i.e., 40°F [4.4°C]). However, when the temperature of the liquid flowing by the sensor becomes higher than the upper set point temperature, the valve will close. The moving liquid mitigates a stagnant flow situation in the air valve that would increase its susceptibility to freezing and also allows warmer liquid to enter the air valve from the pipeline to prevent freezing of the air valve. When freeze protection valves are used, the vault design should incorporate sufficient drainage to accommodate the drainage from the thermostatic valve.

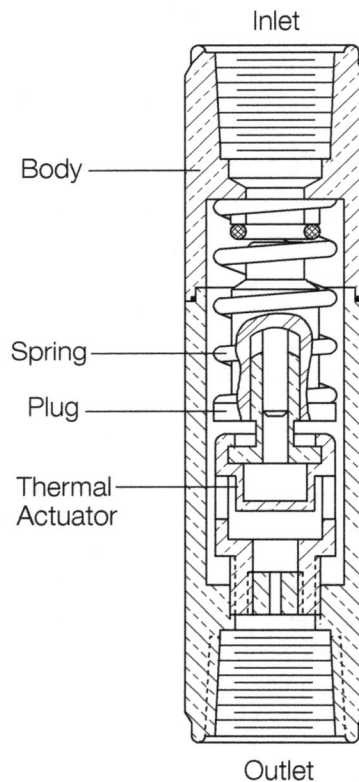

Figure 6-2   Automatic freeze protection valve

Valves Located Belowground

In addition to the protection from freezing and fouling, air valves located belowground should also be provided with a properly vented valve vault. Vaults are subject to local codes, ordinances, and regulations covering cross-connection control and confined-space safety regulations. The vault examples shown in this manual may not meet the requirements of the agency having local jurisdiction.

Air valve vaults come in many different varieties and must all have adequate screened ventilation to satisfy the air requirements for the valve and ventilation of the structure, as shown in Figure 6-3. The two vent pipes are equal to or larger than the air valve size and should provide for the required flow. Valve vaults should be large enough to provide a minimum of 2 ft (0.6 m) of clearance around and above the air valve for maintenance and removal. Intake piping should at minimum include a downturned elbow, an air gap, and a bird screen.

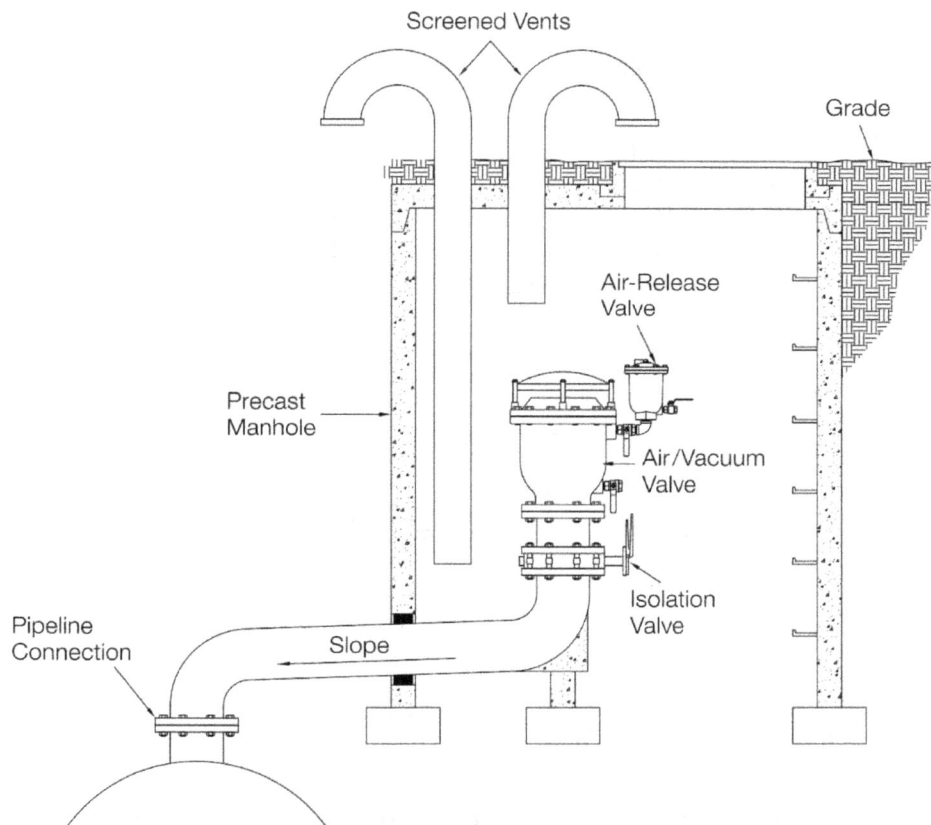

Figure 6-3   Belowground air valve vault installation of a combination air valve (not subject to flooding)

Custom-packaged air valve vaults are also available and are outfitted with special equipment that enables full servicing and maintenance from ground level, including valve removal, without the need of personnel entry into the valve vault (Figure 6-4).

If the vault is subject to flooding, the air valve discharge can be plumbed to a riser above grade with an air gap above the flood plain (Figure 6-5). It is important to note that the riser is exposed to the elements and may be subject to freezing in cold climates.

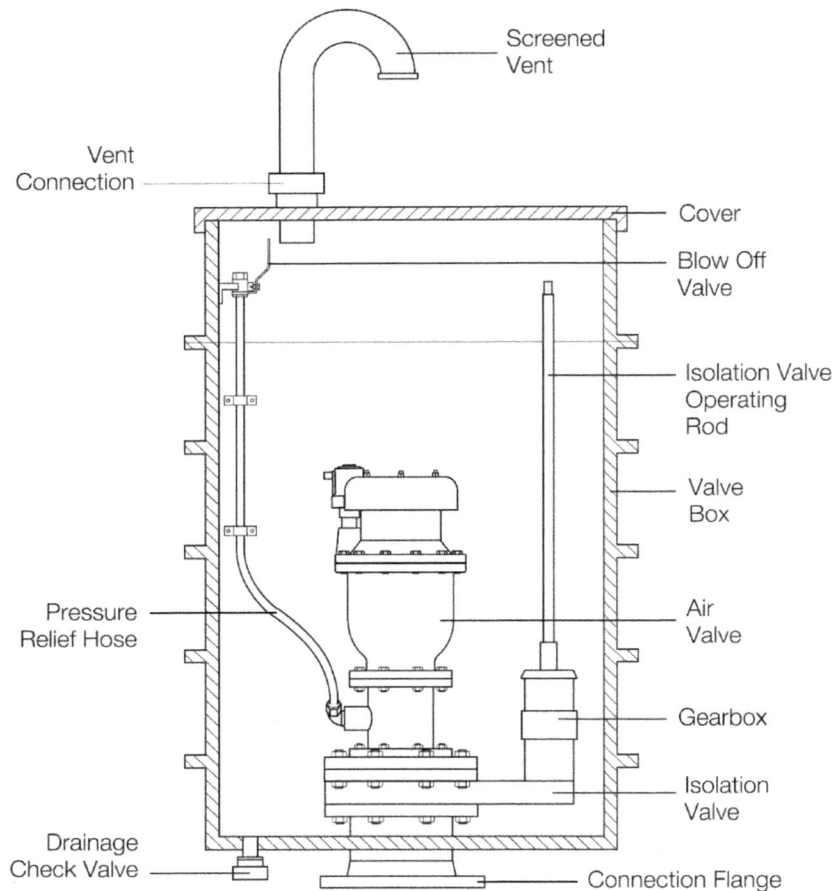

Figure 6-4   Custom vault installation of a combination air valve for servicing aboveground

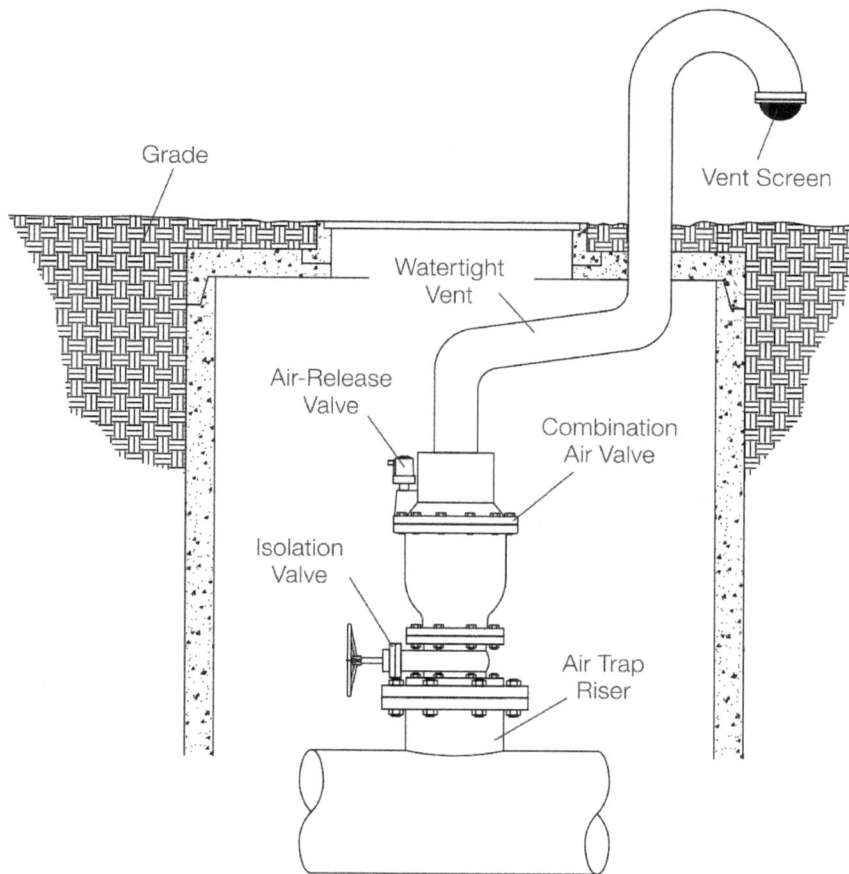

Figure 6-5   Vault installation of a combination air valve with plumbed exhaust

Air Valve Vault Subject to Freezing and Flooding Conditions

In colder climates, an air valve vault can be equipped with two-way dampers to prevent freezing. The vent pipe provides regular airflow but is equipped with two-way dampers to prevent the convection of cold air. When required by the air valve, the dampers open fully to allow flow in both directions. The vent piping is also equipped with screens in two locations to prevent contamination from birds, rodents, and insects and also to prevent intentional tampering by a party with malicious intent.

When a riser pipe is not desirable due to location of the vault (i.e., in a roadway) or there is a risk of flooding or malicious tampering, an inflow preventer built in accordance with ANSI/AWWA C514 can be installed on the valve outlet as shown in Figure 6-6 to prevent contamination. ANSI/AWWA C514 inflow preventers are cross-connection assemblies piped to the outlet of air valves and vents to prevent the entry of contaminated water into the potable water distribution system or storage facilities during flooded conditions. Under normal conditions, the inflow preventer allows air flow in and out of air valves and vents, but when the assembly is submerged, redundant float-operated closure mechanisms automatically close to prevent the entry of contaminated water into the potable water distribution system or storage facilities. The backflow device is field tested annually by certified personnel.

If the vault becomes flooded, the inflow preventer will still allow air to be exhausted by the air valve, while preventing the entry of contaminated water. However, since the entry of water is blocked, no vacuum protection will be afforded while the vault is flooded. If vacuum protection is essential to the structural integrity of the pipeline, then the vault arrangements shown in Figure 6-5 should be employed.

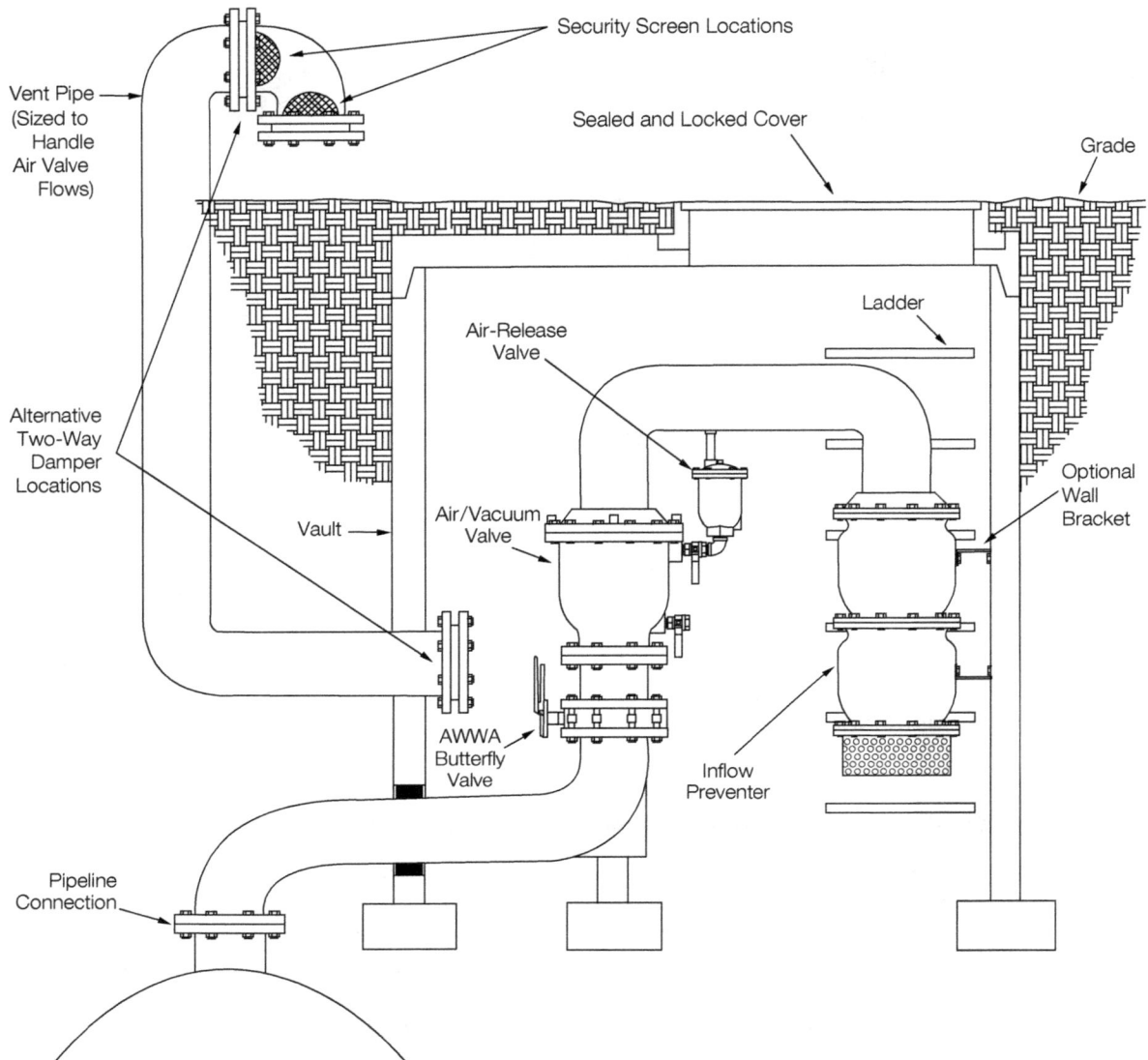

Figure 6-6   Vault installation with freeze and flood protection

OPERATION AND MAINTENANCE

The manufacturers' recommendations on air valve operation and maintenance should be followed.

Continuously Operating Air Valves

Air valves that operate continuously should be opened and flushed more often than valves used for filling and draining. All air valves should be opened and flushed at least annually. Wastewater valves can be furnished with special backwash accessory flushing kits for this purpose.

Filling and Draining Liquid Piping Systems

Caution is required when filling or draining liquid piping systems that have air/vacuum or combination air valves installed on the pipeline (see chapter 5). Never prop and force the valve open by inserting objects into the valve venting port. This can damage the valve's components, and the object may fall into the valve.

Inspection and Maintenance

Air valves should be inspected at least annually for leakage, and the worn parts and resilient seats should be replaced as necessary. A leaky valve is often a result of debris caught in the valve between the float and the resilient seat. Once the hood or outlet piping is removed, the debris can usually be removed without removing the valve cover.

A complete inspection is possible by closing the isolation valve under the air valve, venting the pressure in the valve, and removing the valve cover. WARNING: REMOVING THE COVER OF AN AIR VALVE REQUIRES ISOLATION AND VENTING THE PRESSURE WITHIN THE VALVE OR SERIOUS BODILY HARM MAY RESULT.

Connected to the cover typically are the float and linkage assembly (Figure 2-1). The resilient seats should be checked for damage, cracks, and wear. The float should be checked for the presence of holes, dents, or liquid. The linkage may contain pins and clips that should be checked for structural integrity. Clean or replace parts as needed. A collapsed float may be the result of freezing conditions or pressure surges, which should be addressed before placing the valve back in service. Severe corrosion of parts may be a result of aggressive water or corrosive wastewater. Consult with the manufacturer for the proper materials or coatings for the intended service.

When placing the air valve back in service, verify that all of the valve parts were correctly reassembled in the valve and the cover bolts are tightened in accordance with the valve manufacturer's recommended torques. Open the isolation valve slowly so that the air valve vents air and wastewater gases slowly and fills with liquid over several seconds or a water hammer may occur. Once the valve is pressurized, check the cover for leakage and retighten the cover bolts as needed.

SAFETY

Underground Structures

Hazardous gases collecting in underground structures have caused injuries and fatalities. Gases drawn into a pipe can exit through air valves and remain in the underground structure. Always ventilate the underground structure and use a combustible-gas and low-oxygen detector before entering the structure. Consult Occupational Safety and Health

Administration (OSHA) rules and procedures, such as the need for harnesses and ground-level supervision, in all underground work.

Prepackaged air valves, complete with custom valve vaults (Figure 6-4) outfitted with special equipment that enables full servicing and maintenance from the surface including valve removal, do not require personnel to enter the valve vault and therefore do not require special confined space entry precautions such as the need for harnesses and ground-level supervision.

Inspection of Air Valves

When inspecting air valves, disable the valve by closing the shutoff valve before putting hands and fingers into the valve outlet. If the air valve should suddenly close, serious injury to hands or fingers may occur.

Pressurized air and wastewater gases can also be trapped between the shutoff valve and the air valve; therefore, any removal of air valve bolts, plugs, or covers must be done with extreme care to release trapped air and wastewater gases slowly and prevent serious injury.

If thread protectors and packing material are not removed from air valves prior to filling the liquid piping system, air and wastewater gases may be trapped in the pipeline.

Do not look into a pressurized air valve without adequate eye and face protection.

If the air valve is not venting, open a drain plug to see if there are air and wastewater gases in the valve. An air-release valve may not release air and wastewater gases if the maximum operating pressure is exceeded. Air/vacuum valves only release air and wastewater gases at pressures below atmospheric pressure (i.e., during pipeline filling). Once the system is pressurized, an air/vacuum valve will not open to release air and wastewater gases until the pipeline returns to below atmospheric pressure. Consult the manufacturer for the operating pressure of the installed air valve.

REFERENCES

Van Vuuren, S.J., M. Van Dijk, and J.N. Steenkamp. 2004. *Quantifying the Influence of Air on the Capacity of Large Diameter Water Pipelines and Developing Provisional Guidelines for Effective De-Aeration—Volume 2.* WRC Report No. 1177/2/04. Pretoria, South Africa: Water Research Commission.

Bibliography

ANSI/AWWA C512. Air Release, Air/Vacuum, and Combination Air Valves for Water and Wastewater Service. Denver, CO: AWWA.

ANSI/AWWA C514. Air Valve and Vent Inflow Preventer Assemblies for Potable Water Distribution System and Storage Facilities. Denver, CO: AWWA.

Boulos, P.F., B.W. Karney, D.J. Wood, and S. Lingireddy. 2005. Hydraulic Transient Guidelines for Protecting Water Distribution Systems. *Jour. AWWA*, 97(5):111–124.

Colorado State University. 1977. *Concepts of Water Hammer and Air Entrapment in the Filling and Testing of Pipelines*. Fort Collins, CO: Colorado State University.

Crane Co., Engineering Division. 1982. *Flow of Fluids Through Valves, Fittings and Pipe*. Technical Paper No. 410. New York: Crane Co.

Dean, J.A. 1992. *Lange's Handbook of Chemistry*, 14th ed. New York: McGraw-Hill.

Edmunds, R.C. 1979. Air Binding in Pipes. *Jour. AWWA*, 71(5): 272–277.

Escarameia, M.; C. Dabrowski, C. Gahan, and C. Lauchlan. 2005. *Experimental and Numerical Studies on Movement of Air in Water Pipelines*, Report SR 661, Release 3.0. Wallingford, Oxfordshire, UK: HR Wallingford Ltd.

Giles, R.V. 1962. *Theory and Problems of Fluid Mechanics and Hydraulics*, 2nd ed. (Schaum's Outlines). New York: McGraw-Hill.

Jones, G.M., R.L. Sanks, G. Tchobanoglous, and B.E. Bosserman II, eds. 2008. *Pumping Station Design*, rev. 3rd ed. Boston: Butterworth-Heinemann.

Karassik, I.J., J.P. Messina, P. Cooper, and C.C. Heald. 2008. *Pump Handbook*, 4th ed. New York: McGraw-Hill.

Kroon, J.R., M.A. Stoner, and W.A. Hunt. 1984. Water Hammer: Causes and Effects. *Jour. AWWA*, 76(11):39–45.

Lescovich, J.E. 1972. Locating and Sizing Air-Release Valves. *Jour. AWWA*, 64(7):457–461.

Lubbers, C.L., and F.H.L.R. Clemens. 2005. Capacity Reduction Caused by Air Intake at Wastewater Pumping Stations. In *Proc. Water and Wastewater Pumping Stations Conference, April*. Cranfield, UK: University of Cranfield.

Pozos-Estrada, O. 2007. Investigation on the Effects of Entrained Air in Pipelines, Dissertation Volume 158. Stuttgart, Germany: Institute of Hydraulic Engineering Faculty of Civil and Environmental Engineering, Engineering and Allied Operations, Institut für Wasserbau der Universität Stuttgart.

Tchobanoglous, G., Metcalf and Eddy Inc. 1981, *Wastewater Engineering: Collection and Pumping of Wastewater*. New York: McGraw-Hill.

Thorley, A.R.D. 2004. *Fluid Transients in Pipeline Systems: A Guide to the Control and Suppression of Fluid Transients in Liquids in Closed Conduit*, 2nd ed. New York: ASME.

Tullis, J.P. 1989. *Hydraulics of Pipelines: Pumps, Valves, Cavitation, Transients*. New York: John Wiley and Sons.

Val-Matic Valve and Manufacturing Corp. 2015. *Theory, Application, and Sizing of Air Valves.* Elmhurst, IL: Val-Matic Valve and Manufacturing Corp.

Van Vuuren, S.J., M. Van Dijk, and J.N. Steenkamp. 2004. *Quantifying the Influence of Air on the Capacity of Large Diameter Water Pipelines and Developing Provisional Guidelines for Effective De-Aeration—Volume 2.* WRC Report No. 1177/2/04. Pretoria, South Africa: Water Research Commission.

Wisner, P.E. 1975. Removal of Air From Water Lines. *Journal of the Hydraulics Division, ASCE.* February.

Wood, D.J. Latest edition. *Surge Reference Manual.* Lexington: Department of Civil Engineering, University of Kentucky.

Index

Note: *f.* indicates figure; *t.* indicates table

This page intentionally blank.

AWWA Manuals

M1, *Principles of Water Rates, Fees, and Charges, #30001*

M2, *Instrumentation and Control, #30002*

M3, *Safety Management for Water Utilities, #30003*

M4, *Water Fluoridation Principles and Practices, #30004*

M5, *Water Utility Management, #30005*

M6, *Water Meters—Selection, Installation, Testing, and Maintenance, #30006*

M7, *Problem Organisms in Water: Identification and Treatment, #30007*

M9, *Concrete Pressure Pipe, #30009*

M11, *Steel Pipe—A Guide for Design and Installation, #30011*

M12, *Simplified Procedures for Water Examination, #30012*

M14, *Backflow Prevention and Cross-Connection Control: Recommended Practices, #30014*

M17, *Fire Hydrants: Installation, Field Testing, and Maintenance, #30017*

M19, *Emergency Planning for Water Utilities, #30019*

M20, *Water Chlorination/Chloramination Practices and Principles, #30020*

M21, *Groundwater, #30021*

M22, *Sizing Water Service Lines and Meters, #30022*

M23, *PVC Pipe—Design and Installation, #30023*

M24, *Planning for the Distribution of Reclaimed Water, #30024*

M25, *Flexible-Membrane Covers and Linings for Potable-Water Reservoirs, #30025*

M27, *External Corrosion Control for Infrastructure Sustainability, #30027*

M28, *Rehabilitation of Water Mains, #30028*

M29, *Water Utility Capital Financing, #30029*

M30, *Precoat Filtration, #30030*

M31, *Distribution System Requirements for Fire Protection, #30031*

M32, *Computer Modeling of Water Distribution Systems, #30032*

M33, *Flowmeters in Water Supply, #30033*

M36, *Water Audits and Loss Control Programs, #30036*

M37, *Operational Control of Coagulation and Filtration Processes, #30037*

M38, *Electrodialysis and Electrodialysis Reversal, #30038*

M41, *Ductile-Iron Pipe and Fittings, #30041*

M42, *Steel Water-Storage Tanks, #30042*

M44, *Distribution Valves: Selection, Installation, Field Testing, and Maintenance, #30044*

M45, *Fiberglass Pipe Design, #30045*

M46, *Reverse Osmosis and Nanofiltration, #30046*

M47, *Capital Project Delivery, #30047*

M48, *Waterborne Pathogens, #30048*

M49, *Butterfly Valves: Torque, Head Loss, and Cavitation Analysis, #30049*

M50, *Water Resources Planning, #30050*

M51, *Air Valves: Air-Release, Air/Vacuum and Combination, #30051*

M52, *Water Conservation Programs—A Planning Manual, #30052*

M53, *Microfiltration and Ultrafiltration Membranes for Drinking Water, #30053*

M54, *Developing Rates for Small Systems, #30054*

M55, *PE Pipe—Design and Installation, #30055*

M56, *Nitrification Prevention and Control in Drinking Water, #30056*

M57, *Algae: Source to Treatment, #30057*

M58, *Internal Corrosion Control in Water Distribution Systems, #30058*

M60, *Drought Preparedness and Response, #30060*

M61, *Desalination of Seawater, #30061*

M63, *Aquifer Storage and Recovery, #30063*

M65, *On-Site Generation of Hypochlorite, #30065*

M66, *Cylinder and Vane Actuators and Controls—Design and Installation, #30066*

This page intentionally blank.

This page intentionally blank.

This page intentionally blank.

www.ingramcontent.com/pod-product-compliance
Lightning Source LLC
Chambersburg PA
CBHW081554220326
41598CB00036B/6669